REED CONCISE GUIDE

LIZARDS
of Australia

Steve K. Wilson

First published in 2020 by Reed New Holland Publishers
Sydney

Level 1, 178 Fox Valley Road, Wahroonga, NSW 2076, Australia

www.newhollandpublishers.com

Copyright © 2020 Reed New Holland Publishers
Copyright © 2020 in text and images: Steve K. Wilson or as credited

All rights reserved. No part of this publication may be reproduced, stored in a retrieval system or transmitted, in any form or by any means, electronic, mechanical, photocopying, recording or otherwise, without the prior written permission of the publishers and copyright holders.

A record of this book is held at the National Library of Australia.

All images by Steve K. Wilson except as credited.

ISBN 978 1 92554 657 6

Managing Director: Fiona Schultz
Publisher and Project Editor: Simon Papps
Designer: Yolanda La Gorcé
Production Director: Arlene Gippert
Printed in China

10 9 8 7 6 5 4 3 2

Keep up with Reed New Holland
and New Holland Publishers

f ReedNewHolland

◎ @NewHollandPublishers and @ReedNewHolland

CONTENTS

The land of lizards 5
Species covered in this book 6
Diets 7
Reproduction 9
Temperature management 11
Leaf-tailed, thick-tailed and knob-tailed geckos.
   Family CARPHODACTYLIDAE 13
Austral Geckos. Family DIPLODACTYLIDAE 24
Cosmopolitan Geckos. Family GEKKONIDAE 44
Flap-footed Lizards. Family PYGOPODIDAE 54
Skinks. Family SCINCIDAE 64
Dragons. Family AGAMIDAE 143
Goannas or Monitors. Family VARANIDAE 174
Further reading 186
Glossary 187
Index 188

Top Left: There are few areas in Australia without lizards. Even the central city areas of Sydney and Brisbane support thriving colonies of Water Dragons *(Intellagama lesueurii)*. Brisbane, Qld.

Top Right: Spinifex is a critical provider of habitat in Australian arid areas.

Lizards thrive in tropical and subtropical rainforests but they can be very difficult to find in these habitats.

THE LAND OF LIZARDS

There are more than 800 species of lizards in Australia, and that number grows each year. Lizards live virtually everywhere, from parks and gardens in our towns and cities, to peaks of the highest mountains, and plains of the arid interior. Some thrive in the harshest conditions.

The areas with the greatest species richness of lizards are deserts vegetated with prickly grass hummocks called spinifex. Each hummock, protected by a matrix of sharp spines, forms a shaded microclimate. Many species do not occur anywhere else. Tropical woodlands are also excellent lizard habitats, along with the escarpments and outcrops of Kakadu and the Kimberleys, and forested ranges of Cape York.

The rainforests between north-eastern Queensland's Wet Tropics and subtropical north-eastern New South Wales support unique suites of species, many of them easily observed via accessible walking tracks. The south-west, famous for endemic flowering plants, harbours many reptiles that occur nowhere else.

Diversity is much lower in the Southern Alps and Tasmania. There is less variety but high abundance; the gatherings of basking skinks will virtually part before you on sunny mornings in the alpine tussock grasslands.

Welcome to Australia – the land of lizards!

SPECIES COVERED IN THIS BOOK

It is not possible to feature all of the more than 800 Australian lizard species into this concise guide. Those selected are commonly encountered, spectacular in appearance, of conservation concern, or have unusual habits or stories to tell. Species from all corners of Australia are included, so no matter where the reader lives or travels, they may see lizards featured here. To ensure a broad representation of Australian lizard fauna at least one member from every genus is included.

This book is not a definitive identification guide. Many species featured are extremely similar to others not included. Identification can be complex and difficult. Even experts with all available literature sometimes struggle. Recommended identification books are listed near the end of the book.

The book is designed as a general guide that fits easily into a pocket and helps promote interest in Australian lizards. It is a portable window into Australia's diverse lizard fauna, and hopes to shed light on some of these remarkable animals that call Australia home.

In the accounts abbreviations for measurements are:

SVL for Snout-Vent Length – the standard measurement for a lizard excluding tail; and

TL for Total Length – large species are often measured to include the tail.

DIETS

Most small lizards are opportunistic predators of insects, but with increasing size there is a shift towards an omnivorous diet. The Land Mullet, a giant skink from the subtropical rainforests, consumes large amounts of fungi. Other big skinks such as blue-tongues take fruits, flowers and soft foliage, along with snails, eggs and insects. Large dragons such as bearded dragons follow the same trend, although one large dragon, the Frill-neck, remains carnivorous. Our largest lizards, the goannas or monitors, are exclusively carnivorous, eating carrion, mammals, reptiles, insects and eggs.

This Black-throated Rainbow Skink *(Carlia rostralis)* has captured a grasshopper. Davies Creek, Qld.

Some skinks and geckos occasionally visit flowers for nectar. Many geckos also lick sap oozing from tree trunks. There are specialists too. Beaked geckos feed only on termites, and the Pink-tongued Skink has modified enlarged rear teeth to crush snail shells. Burton's Snake-lizard is a lizard hunter. The snout has backward-curved teeth and the long head is hinged level with the eyes so the tips of the snout join around the prey. The most famous specialist, the Thorny Devil, eats only small black ants. It feeds by positioning itself above ant trails, so that its food effectively self-delivers in an orderly fashion.

Native raspberry fruits make a fine meal for a Major Skink *(Bellatorias frerei)*. Springbrook, Qld.

The Common Scaly-foot *(Pygopus lepidopodus)* is particularly fond of ground spiders. Mt Nebo, Qld.

REPRODUCTION

Most lizards lay eggs deposited in sheltered humid sites such as compost, and in purpose-built burrows. Some Australian skinks give birth to fully formed young, effectively retaining thin-shelled eggs longer in the body.

Cool climates are not always suitable for egg incubation so most of Tasmania's lizards are live-bearers. In the tropics egg-layers are far more common but live-bearing skinks such as blue-tongues live all over Australia.

There are costs and benefits both ways. A live-bearer controls the incubation of her young, but with a heavier burden, she faces a higher predation risk.

Two male Water Dragons *(Intellagama lesueurii)* prepare for a bout of combat. The winner will likely have more mating opportunities. Brisbane, Qld.

Communal egg-laying is common among some lizards. Multiple females deposit their clutches together, opting for an 'all in one basket' strategy. These are Garden Skink *(Lampropholis delicata)* eggs. Gerringong, NSW.

Most lizard eggs have soft pliable shells, prone to desiccation in dry conditions. But some geckos lay eggs with hard, relatively impermeable shells, rather like those of birds.

For most animals, including humans, the sex of an offspring is determined at fertilisation. However for some lizards lower incubation temperatures result in more females and higher temperatures produce more males. This temperature-dependant sex determination is also a feature of crocodiles and sea turtles.

The Mourning Gecko and some populations of the Bynoe's Gecko exist as all-female populations. The condition, called parthenogenesis, means unfertilised eggs produce female clones.

TEMPERATURE MANAGEMENT

Reptiles are often labelled as 'cold blooded'. This misleading term assumes their temperature is lower than our own. Many reptiles actually have a preferred operating temperature much higher than ours. The main difference is that birds and mammals generate their body heat internally (endothermy) while reptiles source theirs externally (ectothermy) by basking or selecting warm substrates. To gain heat, lizards often flatten their bodies, with their backs facing the sun. Some dragons can also change colour, selecting dark hues to gain heat. They reduce heat with paler, heat-reflecting colours, adopting postures that angle them into the sun to reduce exposed surfaces, and raising themselves above a hot substrate.

In searing temperatures an Eyrean Earless Dragon *(Tympanocryptis tetraporophora)* raises its body high off the hot substrate and angles itself into the sun. Winton area, Qld.

In the Australian Alps this Southern Water Skink *(Eulamprus tympanum)* maximises her heat gain by lying with her body flattened and back facing the sun. Falls Creek, Vic.

Lizards can also lose heat through panting, and by moving into the shade or under shelter.

For nocturnal lizards, ambient temperature of a warm evening is normally sufficient. Some can operate at surprisingly low temperatures. On a chilly autumn evening in Queensland's New England Tablelands the Granite Belt Leaf-tailed Geckos still emerge from their crevices to rest on rock faces while the observer shivers in a bracing wind.

CARPHODACTYLIDAE

LEAF-TAILED, THICK-TAILED AND KNOB-TAILED GECKOS Family Carphodactylidae

The Prickly Knob-tailed Gecko (*Nephrurus asper*) has a greatly reduced tail tipped with a ball-like structure.

Large, spectacular geckos sporting an array of highly modified tails: broadly splayed to bulbous or tuber-shaped, sometimes with a spherical ball on the tip. Scales soft, non-glossy, often with scattered tubercles. Eyes large and lidless. They use their broad flat tongues to clean the transparent spectacle covering the eye. Terrestrial, arboreal and rock-inhabiting lizards with simple, clawed digits. None have pads. When harassed they raise their bodies and utter a rasping bark or squeak. Nocturnal. Endemic to Australia

CARPHODACTYLIDAE

Chameleon Gecko *Carphodactylus laevis*

SIZE/ID: SVL 130mm. Laterally compressed with high vertebral ridge. Body scales uniform. Long thin limbs. Digits without pads. Tail carrot shaped. Brown with original tail strikingly banded black and white. Regenerated tail dull brown with streaks.

HABITAT/RANGE: Rainforests of Wet Tropics, north-eastern Qld.

BEHAVIOUR: Mainly terrestrial but often perches with head downwards at night on slender saplings, probably to ambush passing prey. The tail is unique in producing a loud stridulating noise when severed and wriggling.

Lake Eacham, Qld.

CARPHODACTYLIDAE

Centralian Knob-tailed Gecko *Nephrurus amyae*

Alice Springs, NT.

SIZE/ID: SVL 135mm. Large head and rotund body. Body scales granular mixed with large spiny tubercles, each surrounded by rosettes of smaller tubercles. Digits without pads. Tail minute, tipped with round ball. Rich reddish-brown, with pale spots centred on tubercles.

HABITAT/RANGE: Rock outcrops of arid southern NT.

BEHAVIOUR: Feeds on insects and smaller geckos. Purpose of the knob unclear, but vibrated prior to lunging at prey. One of only three geckos in Australia (possibly the world) unable to discard and regrow its tail.

CARPHODACTYLIDAE

Smooth Knob-tailed Gecko *Nephrurus levis*

Barkly region, NT.

SIZE/ID: SVL 102mm. Large head and rotund body. Body scales granular mixed with small scattered tubercles. Digits without pads. Tail plump and heart shaped ending with enlarged knob. Pink to reddish-brown with dark bands across neck and shoulders, and scattered pale spots.

HABITAT/RANGE: Widespread across arid regions, on open sands and loams, often with spinifex.

BEHAVIOUR: Shelters in burrows, often those abandoned by other lizards. If provoked, rears body high on slender limbs, gapes mouth and lunges at aggressor, uttering a wheezing bark.

CARPHODACTYLIDAE

McIlwraith Leaf-tailed Gecko *Orraya occulta*

Peach Creek, McIlwraith Range, Qld.

SIZE/ID: SVL 108mm. Strongly dorsally depressed with long neck. Scales granular mixed with scattered tubercles over body and limbs. Digits without pads. Tail teardrop shaped, edged with spines when original but smooth and rounded when regenerated. Grey with darker mottling.

HABITAT/RANGE: Restricted to McIlwraith Range, Cape York, Qld, inhabiting boulders along creek-line with rainforest.

BEHAVIOUR: Poorly known due to remote locality. Listed as a Vulnerable species.

CARPHODACTYLIDAE

Mount Blackwood Broad-tailed Gecko *Phyllurus isis*

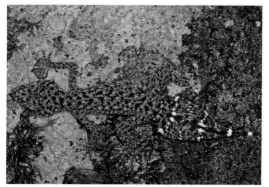

Mt Blackwood, Qld.

SIZE/ID: SVL 76mm. Strongly dorsally depressed. Scales granular, mixed with scattered tubercles over body, limbs and original tail. Digits without pads. Tail broad and flat; spiny with pointed tip on original but smooth and blunt when regenerated. Greyish-brown with lichen-like mottling and pale bands on original tail.

HABITAT/RANGE: Restricted to rainforest with rocks on Mt Blackwood and Mt Jukes, mid-eastern Qld.

BEHAVIOUR: Slow-moving ambush hunter, superbly camouflaged against mossy rocks. Listed as Vulnerable in Qld due to highly restricted distribution.

CARPHODACTYLIDAE

Broad-tailed Gecko *Phyllurus platurus*

Lane Cove National Park, NSW.

SIZE/ID: SVL 95mm. Strongly dorsally depressed. Scales granular, mixed with scattered tubercles over body, limbs and original tail. Digits without pads. Tail broad and flat; spiny with pointed tip on original but smooth and blunt when regenerated. Brown to grey with darker mottling.

HABITAT/RANGE: Sandstone caves and crevices on cliffs and outcrops in Sydney region, NSW, often entering dwellings and garages in close proximity to rocky habitats.

BEHAVIOUR: Large numbers often share crevices, their presence indicated by scattered fragments of shed skin.

CARPHODACTYLIDAE

Northern Leaf-tailed Gecko *Saltuarius cornutus*

SIZE/ID: SVL 144mm. Strongly dorsally depressed. Scales granular, mixed with scattered tubercles over back, limbs and original tail, and recurved hooked tubercles on flanks. Digits without pads. Tail broad and flat; spiny with pointed tip on original but smooth and blunt when regenerated. Grey to brown with lichen-like variegations and blotches.

HABITAT/RANGE: Rainforests of Wet Tropics, north-eastern Qld.

BEHAVIOUR: Mainly arboreal, often dwelling in the latticed trunks of large figs. Slow-moving ambush hunter, emerging at night to rest head downwards on trunks.

Lake Eacham, Qld.

CARPHODACTYLIDAE

Southern Leaf-tailed Gecko *Saltuarius swaini*

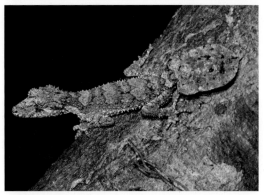

Mt Glorious, Qld.

SIZE/ID: SVL 134mm. Strongly dorsally depressed. Scales granular, mixed with scattered tubercles over body, limbs and original tail. Digits without pads. Tail broad and flat; spiny with pointed tip on original but smooth and blunt when regenerated. Grey to brown with lichen-like variegations and blotches.

HABITAT/RANGE: Subtropical rainforests in the ranges of north-eastern NSW and south-eastern Qld.

BEHAVIOUR: Similar habits to the Northern Leaf-tailed Gecko. Commonly recorded on human dwellings that abut rainforest

CARPHODACTYLIDAE

Common Thick-tailed Gecko *Underwoodisaurus milii*

Karalee Rock, WA.

SIZE/ID: SVL 100mm. Large head, plump body and slender limbs. Scales small and granular mixed with scattered tubercles. Digits without pads. Original tail plump and carrot shaped with tubercles and tapered tip. Regenerated tail smooth, bulbous and blunt-tipped. Pink to purplish-brown with bands of cream spots centred on tubercles. Original tail black and white banded.

HABITAT/RANGE: Well-drained, often rocky habitats across southern Australia.

BEHAVIOUR: Terrestrial, sheltering in burrows and under rocks. When provoked, rears body, lunges and barks with mouth agape.

CARPHODACTYLIDAE

Border Thick-tailed Gecko *Uvidicolus sphyrurus*

Girraween National Park, Qld.

SIZE/ID: SVL 70mm. Large head, plump body and slender limbs. Scales small and granular mixed with scattered tubercles. Digits without pads. Original tail plump and squarish with tubercles and sharply tapered tip. Regenerated tail smooth and blunt tipped. Brownish-grey with dark variegations and pale spots. Original tail black and white banded.

HABITAT/RANGE: Cool rocky highlands on New England Tablelands of northern NSW and southern Qld.

BEHAVIOUR: Terrestrial, sheltering in burrows and under rocks. When provoked, rears body, lunges and barks with mouth agape. Vulnerable species.

AUSTRAL GECKOS Family Diplodactyldae

The Gulf Marbled Velvet Gecko *(Oedura bella)* is an attractive rock-inhabiting species from north-western Queensland and adjacent Northern Territory. Mt Isa area, Qld.

A diverse group. Scales soft, non-glossy, usually smooth but occasionally with tubercles. Eyes large and lidless. They use their broad flat tongues to clean the transparent spectacle covering the eye. Includes terrestrial, arboreal and tussock-inhabiting species. Some have padded toes to climb smooth surfaces of tree trunks and rocks surfaces, while others with greatly reduced pads live on the ground. Endemic to the Australasian region, including New Zealand and New Caledonia.

DIPLODACTYLDAE

Lesueur's Velvet Gecko *Amalosia lesueurii*

SIZE/ID: SVL 80mm. Depressed head, body and tail. Scales very small and uniform. Subdigital lamellae single at base, paired along digit and ending in two enlarged plates. Greyish with ragged paler vertebral blotches.

HABITAT/RANGE: Rock-inhabiting, in eastern NSW and south-eastern Qld.

BEHAVIOUR: The flat body allows access to narrow gaps under rock slabs.

Yalwal Plateau, NSW.

Zigzag Gecko *Amalosia rhombifer*

SIZE/ID: SVL 70mm. Similar to *A. lesueurii*, with broad unbroken pale zigzagging vertebral stripe from nape along original tail.

HABITAT/RANGE: Arboreal, inhabiting dry forests across northern Australia.

BEHAVIOUR: Shelters in hollow limbs and under loose bark.

Tennant Creek, NT.

DIPLODACTYLDAE

Northern Clawless Gecko *Crenadactylus naso*

Kununurra, WA.

SIZE/ID: SVL 31mm. Extremely small with no claws. Subdigital lamellae enlarged ending with two large circular plates. Brown to grey with narrow stripes.

HABITAT/RANGE: Rocky areas across northern Australia between Kimberley region, WA and north-western Qld.

BEHAVIOUR: Lives inside spinifex, and rarely venturing far from shelter.

South-western Clawless Gecko *Crenadactylus ocellatus*

Karalee Rock, WA.

SIZE/ID: SVL 35mm. Similar to *C. naso*. Brown to grey with clustered pale spots, often dark-edged. Sometimes a pair of rusty pale stripes on tail.

HABITAT/RANGE: Dry wooded areas of south-western WA.

BEHAVIOUR: Terrestrial, sheltering under leaf litter, logs and rocks.

DIPLODACTYLDAE

Variable Fat-tailed Gecko *Diplodactylus conspicillatus*

Barkly Station, NT.

SIZE/ID: SVL 62mm. Body scales uniform with mid-dorsal scales not noticeably larger than those on outer back. First scale on upper lip large; all others small and granular. Tail short and fat with alternating transverse rows of large and small scales. Brown to reddish-brown with light and dark chequered pattern, pale spots and pale streak on snout. Several similar species.

HABITAT/RANGE: Arid interior, mainly hard soils.

BEHAVIOUR: Shelters in vertical spider holes, blocking shaft with tail. Eats only termites.

DIPLODACTYLDAE

Mesa Gecko *Diplodactylus galeatus*

Alice Springs, NT.

SIZE/ID: SVL 55mm. Body scales uniform. Small pads on digits, with two rows of subdigital lamellae ending in a pair of enlarged plates. Brown to reddish-brown with 6–9 prominent, dark-edged pale dorsal blotches on body and original tail. Top of head pale, edged by curved line from eye to eye.

HABITAT/RANGE: Rocky ranges and outcrops of Central Australia.

BEHAVIOUR: Shelters under rocks in mild weather, and in deeper insulated cavities during high temperatures.

DIPLODACTYLDAE

Pretty Gecko *Diplodactylus pulcher*

Charles Darwin Nature Reserve, WA.

SIZE/ID: SVL 62mm. Body scales uniform. Small pads on digits, with granular subdigital lamellae ending in a pair of enlarged plates. Brown to reddish-brown with dark-edged pale dorsal blotches, sometimes merged to form a simple straight-edged stripe, and series of pale spots on flanks.

HABITAT/RANGE: Arid to semi-arid mid-west coast and southern interior of WA, in shrublands on heavy to sandy soils.

BEHAVIOUR: Terrestrial. Shelters under surface debris and in vertical spider holes.

DIPLODACTYLDAE

Eastern Stone Gecko *Diplodactylus vittatus*

Warrawee Station, Qld.

SIZE/ID: SVL 62mm. Body scales uniform. Small pads on digits, with subdigital lamellae in large single series ending in a pair of enlarged plates. Tail relatively short and thick. Brown to grey with pale zigzagging vertebral stripe, sometimes broken into irregular blotches.

HABITAT/RANGE: Very widespread in eastern Australia, in shrublands and open forests.

BEHAVIOUR: Terrestrial, but often perches at night on fallen sticks, presumably to ambush passing invertebrates. Shelters under surface debris and in vertical insect and spider holes.

DIPLODACTYLDAE

Western Velvet Gecko *Hesperoedura reticulata*

Bolgart area, WA.

SIZE/ID: SVL 70mm. Weakly dorsally depressed. Body scales uniform, with dorsal scales minute. Prominent pads on digits, with subdigital lamellae single at base, divided along digit and ending in a pair of enlarged plates. Tail relatively long and fleshy. Brown to grey with broad, dark-edged pale dorsal zone, often including a narrow dark vertebral and transverse lines.

HABITAT/RANGE: Woodlands in southern interior of WA.

BEHAVIOUR: Arboreal, sheltering in tree hollows, particularly on smooth-barked eucalypts.

DIPLODACTYLDAE

Beaded Gecko *Lucasium damaeum*

Hattah-Kulkyne National Park, Vic.

SIZE/ID: SVL 55mm. No pads on digits; all subdigital lamellae minute. Pale reddish-brown with paler vertebral stripe or blotches, and large pale lateral spots.

HABITAT/RANGE: Arid areas with spinifex in southern and central Australia.

BEHAVIOUR: Terrestrial, sheltering in vertical shafts of insect and spider holes.

Sand-plain Gecko *Lucasium stenodactylus*

SIZE/ID: SVL 57mm. Similar, but with two minute terminal pads. Reddish-brown with paler vertebral stripe or blotches. Probably a species complex.

Barkly region, NT.

HABITAT/RANGE: Arid sandy areas across northern and central Australia.

BEHAVIOUR: Shelters in shafts of insect and spider holes; sometimes under surface debris.

DIPLODACTYLDAE

Robust Velvet Gecko *Nebulifera robusta*

Kurwongbah, Qld.

SIZE/ID: SVL 80mm. Dorsally depressed. Body scales uniform, with dorsal scales minute and much smaller than ventrals. Tail thick and fleshy. Prominent pads on digits, with subdigital lamellae single at base, divided along digit and ending in a pair of enlarged plates. Brown to blackish-brown with large squarish pale blotches along back and original tail.

HABITAT/RANGE: Woodlands of eastern Australia.

BEHAVIOUR: Mainly arboreal, sheltering in hollow timber, but often living on walls of human dwellings.

DIPLODACTYLDAE

Inland Marbled Velvet Gecko *Oedura cincta*

Morven, Qld.

SIZE/ID: SVL 108mm. Dorsally depressed. Body scales uniform, with dorsal scales large and granular. Prominent pads on digits, with subdigital lamellae single at base, divided along digit and ending with pair of enlarged plates. Juvenile with simple dark and cream bands, fragmenting and mixed with pale spots on adult. Dark stripe across nape between eyes.

HABITAT/RANGE: Woodlands of eastern interior, and outcrops in Central Australia.

BEHAVIOUR: Arboreal in eastern woodlands and rock-inhabiting on central Australian outcrops and gorges.

DIPLODACTYLDAE

Fringe-toed Velvet Gecko *Oedura felicipoda*

Mitchell Plateau, WA.

SIZE/ID: SVL 105mm. Dorsally depressed. Body scales uniform, with dorsal scales large and granular. Tail short and broad. Prominent pads on digits, with subdigital lamellae forming a fringe along each digit; single at base, divided along digit and ending in a pair of enlarged plates. Juvenile with simple dark and cream bands, fragmenting and mixed with pale spots on adult.

HABITAT/RANGE: Sandstone caves and cliffs of north-western Kimberley region, WA.

BEHAVIOUR: Rock-inhabiting, moving with ease on vertical surfaces and overhangs.

DIPLODACTYLDAE

Southern Spotted Velvet Gecko *Oedura tryoni*

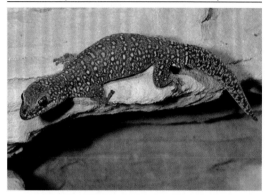

Pony Hills Station, Qld.

SIZE/ID: SVL 87mm. Dorsally-depressed. Body scales uniform, with dorsal scales large and granular. Prominent pads on digits, with subdigital lamellae single at base, divided along digit and ending with pair of enlarged plates. Yellowish-brown with numerous prominent dark-edged pale spots.

HABITAT/RANGE: Woodlands and rock outcrops of north-eastern NSW and south-eastern to mid-eastern Qld.

BEHAVIOUR: Normally rock-inhabiting, typically on granite or sandstone, sheltering in crevices and under slabs. In southern interior of Qld, some arboreal populations live in hollows.

DIPLODACTYLDAE

Giant Tree Gecko *Pseudothecadactylus australis*

Prince of Wales Island, Qld.

SIZE/ID: SVL 120mm. Very large with smooth uniform body scales and cylindrical prehensile tail tipped below with adhesive lamellae. Digits broad and padded with divided series of lamellae along length. Brown to grey with about six pale dorsal blotches often broken into pairs.

HABITAT/RANGE: Vine thickets, rainforest, paperbark woodlands and mangroves of far northern Cape York and southern Torres Strait islands.

BEHAVIOUR: Arboreal, sheltering in hollows. Sometimes betrays its presence by barking if approached. Gapes mouth revealing dark interior if provoked.

DIPLODACTYLDAE

Northern Giant Cave Gecko
Pseudothecadactylus lindneri

Arnhem escarpment, NT.

SIZE/ID: SVL 96mm. Smooth uniform body scales and cylindrical prehensile tail tipped below with adhesive lamellae. Digits broad and padded with lamellae in divided series along length. Brown to dark purplish-brown with irregular pale bands, often off-set along vertebral region. Tail banded.

HABITAT/RANGE: Escarpments and caves and associated vine thickets of western Arnhem Land, NT.

BEHAVIOUR: Mainly rock-inhabiting, foraging on rock faces and ceilings of caves, and onto the branches of adjacent trees such as figs.

DIPLODACTYLDAE

Eyre Basin Beaked Gecko *Rhynchoedura eyrensis*

SIZE/ID: SVL 51mm. Slender cylindrical body, beak-like snout and short limbs. One enlarged mental scale. Dark iris. Reddish-brown with dark-edged pale spots. Several similar species.

HABITAT/RANGE: Shrublands, often with spinifex, in arid eastern interior.

Ethabuka, Qld.

BEHAVIOUR: Shelters in vertical spider holes. Probably a termite specialist.

Brigalow Beaked Gecko *Rhynchoedura mentalis*

SIZE/ID: SVL 50mm. Similar to *R. eyrensis*, with three enlarged mental scales. Iris pale. Dark brown with darker variegations and pale spots or blotches.

HABITAT/RANGE: Shrublands on hard stony soils in arid central Qld.

Quilpie area, Qld.

BEHAVIOUR: Terrestrial. Shelters under small stones and in vertical spider holes. Probably a termite specialist.

DIPLODACTYLDAE

Northern Spiny-tailed Gecko *Strophurus ciliaris*

Barkly region, NT, Qld.

SIZE/ID: SVL 89mm. Long spines over eye. Body scales granular mixed with scattered large tubercles on back, and two impressive rows of spines along tail. Mouth-lining yellow. Variable, from almost white to grey or orange, sometimes blotched.

HABITAT/RANGE: Very widespread over dry areas, from spinifex deserts to tropical woodlands.

BEHAVIOUR: Mainly arboreal, sheltering in hollows, under bark or clinging to exposed branches. If provoked, can squirt sticky irritant fluid from glands along the back and tail, sometime forcibly ejecting it for over a metre.

DIPLODACTYLDAE

Jewelled Gecko *Strophurus elderi*

Ballera area, Qld.

SIZE/ID: SVL 48mm. Short tail. Body scales granular mixed with enlarged tubercles. Toes padded, with subdigital lamellae single at base, divided along digit, ending with pair of enlarged plates. Leaden grey with prominent scattered white spots, each centred on a tubercle.

HABITAT/RANGE: Sandy deserts with spinifex covering most of interior to north-west coast.

BEHAVIOUR: Lives exclusively in spinifex hummocks. If provoked, it squirts irritant fluid from pores in back and tail.

DIPLODACTYLDAE

Southern Phasmid Gecko *Strophurus jeanae*

Barkly region, NT, Qld.

SIZE/ID: SVL 49mm. Extremely slender with long thin limbs and uniform body scales. Toes padded, with subdigital lamellae single at base, divided along digit, ending with pair of enlarged plates. Simply patterned with broad, bold, sharp-edged stripes. Similar species.

HABITAT/RANGE: Sandy deserts with spinifex from central NT to north-west coast.

BEHAVIOUR: Lives exclusively in spinifex hummocks where striped pattern offers superb camouflage. If provoked, it squirts irritant fluid from pores in back and tail.

DIPLODACTYLDAE

Golden-tailed Gecko *Strophurus taenicauda*

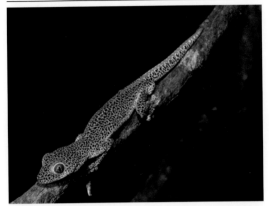

Kumbarilla State Forest, Qld.

SIZE/ID: SVL 73mm. Long body, long cylindrical tail and uniform body scales. Toes padded, with subdigital lamellae single at base, divided along digit, ending with pair of enlarged plates. Pale grey with mosaic of black spots, an orange dorsal blaze on tail and bright red iris. Variable. One subspecies has additional narrow orange stripes along sides of tail. Another has grey iris.

HABITAT/RANGE: Dry forests in eastern interior of Qld.

BEHAVIOUR: Arboreal. If provoked, it squirts irritant fluid from pores in back and tail.

COSMOPOLITAN GECKOS Family Gekkonidae

The Purnululu Dtella *(Gehyra ipsa)* inhabits the domed outcrops of the spectacular Bungle Bungles, WA.

Scales soft, non-glossy, smooth or mixed with tubercles. Eyes large and lidless. Broad flat tongues used to clean the transparent spectacle covering the eye. Many can scuttle with ease over smooth walls and even ceilings, thanks to microscopic branched structures called setae beneath expanded toes. Others, with simple claws, live on the ground. Worldwide group with more than 60 named Australian species. Eggs unique in having hard, brittle shells like those of birds, rather than soft, pliable shells produced by other egg-laying lizards. Nocturnal.

GEKKONIDAE

Marbled Gecko *Christinus marmoratus*

Kangaroo Island, SA.

SIZE/ID: SVL 70mm. Long, fleshy tail and uniform body scales. Toes padded with subdigital lamellae in single series ending with pair of enlarged plates, and small non-overlapping scales under tail. Grey with irregular dark marbling. Sometimes a series of reddish to yellow dorsal blotches on tail, particularly on juvenile. Similar species in Nullarbor region.

HABITAT/RANGE: Woodlands, rock outcrops and some urban regions across southern Australia.

BEHAVIOUR: Arboreal and rock-inhabiting. Australia's most southerly gecko.

GEKKONIDAE

Cooktown Ring-tailed Gecko
Cyrtodactylus tuberculatus

Black Mountain, Qld.

SIZE/ID: SVL 120mm. Extremely large with long slender tail and slender clawed toes without pads. Body scales small and granular mixed with longitudinal rows of large tubercles. Boldly marked with cream and dark-edged brown bands on body, dark mottling on top of head and black and white rings on tail. Other similar species on Cape York.

HABITAT/RANGE: Rocky areas in southern Cape York, Queensland.

BEHAVIOUR: Rock-inhabiting on boulders and rock-faces, extending onto adjacent vegetation and occasionally human dwellings. One of Australia's largest geckos.

GEKKONIDAE

Dubious Dtella *Gehyra dubia*

SIZE/ID: SVL 65mm. Toes with circular pads; claw arising from upper surface. Inner digits clawless. Subdigital lamellae single along digit; broad and undivided under pad. Grey with irregular pale spots and irregular dark streaks. Pale without pattern when foraging. Numerous similar species.

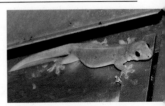

Ganadero Station, Qld.

HABITAT/RANGE: Eastern Australia, in woodlands, rocks and towns.

BEHAVIOUR: Arboreal and rock-inhabiting, often entering buildings.

Northern Spotted Rock Dtella *Gehyra nana*

SIZE/ID: SVL 55mm. Similar to *G. dubia,* with subdigital lamellae divided under pad. Tan to reddish-brown with small pale spots and separated larger dark spots, tending to form transverse rows. Numerous similar species.

HABITAT/RANGE: Rocky areas of northern Australia

BEHAVIOUR: Rock-inhabiting.

El Questro Station, WA.

GEKKONIDAE

Variegated Dtella *Gehyra variegata*

SIZE/ID: SVL 49mm. Toes with large circular pads, with claw arising from upper surface of pad. Inner digits clawless. Subdigital lamellae single along digit and divided under pad. Brown to grey with dark transverse and longitudinal streaks, often with pale spots on rear edges. Numerous similar species.

HABITAT/RANGE: Lower half of WA, in woodlands, rocks and dwellings.

Fraser Range, WA.

BEHAVIOUR: Arboreal and rock-inhabiting, often entering buildings.

Crocodile-faced Dtella *Gehyra xenopus*

SIZE/ID: SVL 77mm. Much larger than *G. variegata*. Flat head and upturned snout. Divided subdigital lamellae under expanded portion of toe, separated at base by wedge of granular scales. Grey with bands of large pale spots.

HABITAT/RANGE: Sandstone outcrops of north-western Kimberley region, WA.

BEHAVIOUR: Rock-inhabiting.

Mitchell Plateau, WA.

GEKKONIDAE

Asian House Gecko *Hemidactylus frenatus*

Kurwongbah, Qld.

SIZE/ID: SVL 60mm. Body scales small and granular mixed with enlarged tubercles; scattered on back and in well-spaced rings on original tail. Digits padded, with subdigital lamellae in divided series. Variable, from grey with darker streaks during the day to cream or pink without pattern when active at night.

HABITAT/RANGE: Towns and cities across north from mid-coastal NSW to mid-west of WA. Usually associated with human structures.

BEHAVIOUR: Introduced with cargo, invasive and rapidly dispersing. Common house gecko with distinctive 'chuck...chuck' call.

GEKKONIDAE

Bynoe's Gecko *Heteronotia binoei*

Fraser Range, WA.

SIZE/ID: SVL 54mm. Body scales small and granular with scattered to longitudinal rows of enlarged tubercles. Toes narrow without pads. Claws surrounded by three scales, one above and two below. Extremely variable from reddish-brown to black with spots and/or ragged bands. Probably a complex of numerous species.

HABITAT/RANGE: Australia-wide except lower south-east and south-west. Most widespread Australian gecko.

BEHAVIOUR: Terrestrial, hiding under any available debris. Usually abundant. Some all-female populations, which reproduce without mating, occur alongside typical bisexual populations.

GEKKONIDAE

Mourning Gecko *Lepidodactylus lugubris*

Cooktown, Qld.

SIZE/ID: SVL 50mm. Body slightly flattened. Tail flattened, edged with fringe of fine scales. Toes padded. Subdigital lamellae single across base of digit, divided along last half. Cream to brown, with dark flecks and often a series of W-shaped markings on back.

HABITAT/RANGE: Mainly human dwellings in northern towns, south to Sunshine Coast. Possibly introduced.

BEHAVIOUR: Arboreal and on walls and fences. All-female, reproducing female clones without mating. Widespread colonist of far-flung Pacific islands.

GEKKONIDAE

Southern Cape York Gecko *Nactus cheverti*

Shipton's Flat, Qld.

SIZE/ID: SVL 57mm. Body scales small and granular with about 20 or fewer longitudinal rows of enlarged conical tubercles. Toes narrow without pads. Claws surrounded by two scales, one above and one below. Brown with pattern (when present) comprising bands or transverse blotches. Similar species further north.

HABITAT/RANGE: Woodlands and outcrops of southern Cape York, from Cairns to Cape Melville.

BEHAVIOUR: Terrestrial, hiding under any available debris.

GEKKONIDAE

Black Mountain Gecko *Nactus galgajuga*

Black Mountain, Qld.

SIZE/ID: SVL 50mm. Slender with large eyes and long thin limbs. Body scales small and granular with longitudinal rows of enlarged conical tubercles. Toes narrow without pads. Claws surrounded by two scales, one above and one below. Purplish-brown with irregular pale bands.

HABITAT/RANGE: Endemic to piled dark boulders of Black Mountain near Cooktown.

BEHAVIOUR: Extremely swift and agile, foraging over boulder surfaces and leaping between rocks. Listed as Vulnerable in Qld.

FLAP-FOOTED LIZARDS Family Pygopodidae

The Javelin Delma *(Delma concinna)* is an extremely swift inhabitant of south-western heaths. Mt Leseur National Park, WA.

These lizards appear limbless and are often mistaken for snakes. They lack fore-limbs, and the tiny hind-limbs are represented as scaly flaps. Flap-footed lizards are unlikely cousins of geckos, sharing lidless eyes, the use of the tongue to clean the eye spectacle, and a voice in the form of a squeak. The 45 Australian species live over most of the continent excluding Tasmania. One species is endemic to New Guinea. They are common in dry areas including spinifex deserts, heaths and grassy woodlands and generally avoid moist habitats.

PYGOPODIDAE

Red-tailed Worm-lizard *Aprasia inaurita*

SIZE/ID: SVL 136mm. Worm-like with blunt-tipped tail, rounded, slightly protrusive snout and minute limb flaps. First upper labial scale partly fused to nasal scale. Brown with bright reddish-orange tail.

HABITAT/RANGE: Mallee woodlands in semi-arid southern Australia.

Port Germein, SA.

BEHAVIOUR: Burrower, often associated with ant nests, preying on the eggs and pupae.

Granite Worm-lizard *Aprasia pulchella*

SIZE/ID: SVL 120mm. Similar to *A. inaurita*, with first upper labial scale fully fused to nasal scale. Brown with little pattern.

HABITAT/RANGE: Darling Range, south-western WA, in on granitic and lateritic soils.

BEHAVIOUR: Similar to *A. inaurita*.

Bungendore, WA.

PYGOPODIDAE

Marble-faced Delma *Delma australis*

Hattah-Kulkyne National Park, Vic.

SIZE/ID: SVL 93mm. Hind-limb flaps moderately well-developed. Tail a little less than 2.5 times length of body. Ventral scales not significantly larger than other body scales. Brown with irregular dark variegations or bars on head, neck and fore-body including chin and throat.

HABITAT/RANGE: Dry to arid southern Australia, in shrublands and mallee woodlands.

BEHAVIOUR: Shelters in spinifex and under thick mats of leaf litter at bases of trees and shrubs.

PYGOPODIDAE

Striped Delma *Delma impar*

Western Melbourne area, Vic.

SIZE/ID: SVL 100mm. Hind-limb flaps moderately well-developed. Tail about 2.5–3 times body length. Ventral scales paired, significantly larger than other body scales. Prominently striped pattern including broad brown vertebral stripe and narrow black lateral lines breaking into oblique bars on tail.

HABITAT/RANGE: Diminishing and fragmented temperate southeastern grasslands in Vic, ACT and NSW..

BEHAVIOUR: Shelters in tussocks and under rocks. Agricultural and urban development are major threats. Declining and listed as Endangered in Vic and Vulnerable in NSW and ACT.

PYGOPODIDAE

Black-necked Delma *Delma tincta*

Barkly region, NT.

SIZE/ID: SVL 92mm. Hind-limb flaps moderately well-developed. Tail about four times body length. Third upper labial scale below eye. Midbody scales in 14 rows. Brown with 3–4 black bands with sharp pale interspaces across head and neck. Pattern fades with age. Other similar species.

HABITAT/RANGE: Widespread in dry habitats over northern Australia.

BEHAVIOUR: Shelters in spinifex, other grasses, leaf litter and under rocks and logs.

Collared Delma *Delma torquata*

SIZE/ID: SVL 63mm. Similar to *D. tincta* with midbody scales in 16 rows, black marbling on throat and pattern present at all ages.

HABITAT/RANGE: Rocky slopes with native tussocks in south-eastern Qld.

BEHAVIOUR: Shelters under small rocks and leaf litter. Listed as Vulnerable.

Withcott area, Qld.

PYGOPODIDAE

Burton's Snake-lizard *Lialis burtonis*

Beerwah, Qld.

SIZE/ID: SVL 290mm. Distinctive pointed, wedge-shaped snout. Hind-limb flaps minute. Easily recognised form but extremely variable colours including pale grey, yellow, reddish-brown with pattern absent, or comprising bold narrow stripes along body, or black and white stripes from face to fore-body.

HABITAT/RANGE: Australia-wide except moist south-east and south-west. Most widespread lizard, also reaching New Guinea, in virtually all dry habitats.

BEHAVIOUR: Specialised diet of lizards, snatched in narrow jaws with hinged teeth and swallowed head-first. Ecologically similar to small snakes.

PYGOPODIDAE

Bronzeback *Ophidiocephalus taeniatus*

Coober Pedy area, SA. [Photo credit. B. Schembri]

SIZE/ID: SVL 102mm. Protrusive snout and small hind-limb flaps. Midbody scales in 16 rows. Ventral scales not significantly larger than other body scales. Tail about 1.5 times body length. Bronze-brown above and grey on flanks, sharply delineated by dark upper lateral line.

HABITAT/RANGE: Arid sparsely vegetated areas of northern SA and southern NT, sheltering under leaf litter beneath trees and shrubs.

BEHAVIOUR: Sand swimmer in upper soil layers, retreating down cracks if disturbed.

PYGOPODIDAE

Brigalow Scaly-foot *Paradelma orientalis*

Strathblane Station, Qld.

SIZE/ID: SVL 197mm. Round snout and moderately large hind-limb flaps. Midbody scales in 18 rows. Ventral scales significantly larger than other body scales. Tail about two times body length. Opaque glossy greyish-brown with cream base of head, darker snout and dark bar across neck.

HABITAT/RANGE: Dry forests in eastern interior of Qld including brigalow, native pine and sandstone ridges. Shelters under rocks, logs and leaf litter.

BEHAVIOUR: Nocturnal. When alarmed, rears head and flickers tongue like small venomous snake.

PYGOPODIDAE

Keeled Legless Lizard *Pletholax gracilis*

Perth, WA. [Photo credit. B. Schembri]

SIZE/ID: SVL 90mm. Extremely slender with pointed snout, small hind-limb flaps and strong keels on all scales. Tail nearly 3.5 times body length. Grey with vertebral and lateral stripes and yellow throat.

HABITAT/RANGE: Sandplains with heath and Banksia along lower west coast, with similar larger species further north in vicinity of Shark Bay.

BEHAVIOUR: Extremely secretive, sheltering and foraging within tussocks and other thick low vegetation.

PYGOPODIDAE

Common Scaly-foot *Pygopus lepidopodus*

SIZE/ID: SVL 274mm. Large and robust with round snout, keeled dorsal scales and very large hind-limb flaps. Uniform reddish-brown; or grey with three prominent rows of white-edged black blotches.

Lamington National Park, Qld.

HABITAT/RANGE: Southern Australia, in heaths or in woodlands over tussock-dominated ground cover.

BEHAVIOUR: Boldly marked lizards occupy heaths. Nocturnal and diurnal. When alarmed, rears head and mimics small venomous snake.

Western Hooded Scaly-foot *Pygopus nigriceps*

SIZE/ID: SVL 227mm. Similar to above, with smooth dorsal scales. Broad dark band across neck and bar through eye to lips. Reddish-brown with usually weak pattern including oblique dark lines. Other similar species.

Barkly region, NT.

HABITAT/RANGE: Central and western deserts. Shelters in soil cracks and cavities.

BEHAVIOUR: Nocturnal. Eats mainly terrestrial spiders.

SKINKS Family Scincidae

With more than 450 species, skinks are the most diverse and widespread Australian lizards.

Typical skinks are small, smooth, shiny lizards with four limbs, overlapping scales and long tails that break easily. Others burrow into loose sand or compost. These have shorter limbs, some lack front legs and digits, and several are limbless.

Some have spiny scales to wedge themselves into crevices, or splayed limbs to shin over vertical surfaces. Most skinks eat small insects but the largest, blue-tongues, the Shingleback and Land Mullet, are omnivorous.

The Metallic Skink *(Carinascincus metallicus)* is common in southern Victoria and Tasmania. Yarram, Vic.

The Chillagoe Fine-lined Slider *(Lerista parameles)* is a burrower with reduced legs. Almaden, Qld.

SCINCIDAE

Eastern Three-lined Skink *Acritoscincus duperryi*

Sorrento, Vic.

SIZE/ID: SVL 80mm. Limbs with five digits. Frontoparietal scales fused into one shield. Moveable lower eyelid enclosing transparent disc. Silver to brown with dark vertebral, dorsolateral and broad upper lateral stripes, white midlateral stripe and reddish throat, brightest on breeding ♂.

HABITAT/RANGE: South-eastern mainland and Tas., in cool temperate forests and heaths, often with tussock ground cover.

BEHAVIOUR: Terrestrial. Diurnal, basking in sheltered sites near cover. Egg-layer, often depositing communally.

SCINCIDAE

McCoy's Skink *Anepischetosia maccoyi*

Olinda, Vic.

SIZE/ID: SVL 50mm. Small with short, widely-spaced limbs with five digits and glossy scales. Brown with narrow dark dorsolateral line and cream to yellow or orange belly.

HABITAT/RANGE: South-eastern mainland, north to south coast of NSW. Moist gullies and other damp sites, sheltering under rotting logs and leaf litter.

BEHAVIOUR: Secretive, dwelling in shaded sites and avoiding sunlight. Feeds on small invertebrates. Egg-layer.

SCINCIDAE

Short-necked Worm Skink *Anomalopus brevicollis*

Cracow, Qld.

SIZE/ID: SVL 83mm. Completely limbless, with waxy-tipped snout, reduced eyes and no ear-opening. Brown to yellowish-brown with darker tail. Pattern largely absent; darker base on each scale.

HABITAT/RANGE: Woodlands, dry forests and outcrops in mid-eastern Qld and interior, inland to Clermont area.

BEHAVIOUR: Burrower, dwelling in thick compost around the bases of trees and in soft soil under rocks and logs.

SCINCIDAE

Verreaux's Skink *Anomalopus verreauxii*

Mackay, Qld.

SIZE/ID: SVL 185mm. Long body with very small limbs. Fore-limbs with three fingers; hind-limbs reduced to stumps. Ear-opening absent. Glossy greyish-brown. Juvenile has prominent yellow bar across nape, darkening and becoming obscure on adult.

HABITAT/RANGE: Forests and woodlands and some suburban areas from south-eastern to mid-eastern Qld.

BEHAVIOUR: Burrows in loose soil under rocks and logs. Egg-layer.

SCINCIDAE

Kinghorn's Snake-eyed Skink
Austroablepharus kinghorni

Durham Downs Station, Qld.

SIZE/ID: SVL 45mm. Slender limbs with five digits. Frontoparietal scales fused into one shield. Eyelid immoveable, a large fixed spectacle. Brownish with numerous narrow dark stripes to hips and a bright reddish-orange tail.

HABITAT/RANGE: Open grasslands on cracking clay and red loams in arid to semi-arid eastern interior.

BEHAVIOUR: Diurnal. Extremely secretive sheltering under tussocks and in soil cracks. Egg-layer.

SCINCIDAE

Major Skink *Bellatorias frerei*

SIZE/ID: SVL 180mm. Extremely large and robust with parietal scales separated. Eyelids pale. Brown with dark streaks above, flanks darker with pale spots.

HABITAT/RANGE: Woodlands, forests and rainforest margins in tropical and subtropical east.

BEHAVIOUR: Diurnal. Dashes noisily to cover if disturbed. Live-bearer.

Cooloola National Park, Qld.

Land Mullet *Bellatorias major*

SIZE/ID: SVL 300mm. Similar to *B. frerei* but larger. Adult glossy black. Juvenile has pale spots.

HABITAT/RANGE: Subtropical rainforests in south-eastern Qld and north-eastern NSW.

BEHAVIOUR: Diurnal. Basks in sheltered sunny patches. Live-bearer.

Springbrook, Qld.

SCINCIDAE

Garden Calyptotis *Calyptotis scutirostrum*

Kurwongbah, Qld.

SIZE/ID: SVL 55mm. Long body. Short, widely-spaced limbs with five digits. Movable scaly lower eyelid. No ear-openings and no prefrontal scales. Glossy brown with four narrow dark dorsal lines, black dorsolateral stripe. Belly yellow with pink flush under tail. Other similar species.

HABITAT/RANGE: Forests, woodlands and suburban gardens in south-eastern Qld and north-eastern NSW.

BEHAVIOUR: Lives under compost and leaf litter, rarely basking. Egg-layer.

SCINCIDAE

Ocellated Skink *Carinascincus ocellatus*

Freycinet National Park, Tas.

SIZE/ID: SVL 85mm. Dorsally depressed. Well-developed limbs with five digits. Frontoparietal scales fused to form one shield. Moveable lower eyelid enclosing transparent disc. Brown to copper with numerous prominent dark-edged pale blotches.

HABITAT/RANGE: Cool temperate rocky areas in open habitats across Tas.

BEHAVIOUR: Mainly rock-inhabiting. Basks on rock faces and logs. Flattened body allows access to narrow crevices. Live-bearer.

SCINCIDAE

Closed-litter Rainbow Skink *Carlia longipes*

SIZE/ID: SVL 68mm. Limbs well-developed; four fingers and five toes. Ear-opening surrounded by sharp lobules. Smooth dorsal scales. ♂ has rich orange flanks. ♀ has pale stripes; under eye to ear; dorsolaterals; and midlaterals. Numerous similar species.

Male. Cooktown, Qld.

HABITAT/RANGE: Coastal north-eastern Qld.

BEHAVIOUR: Sun-loving. Basks in leaf litter.

Blue-throated Rainbow Skink *Carlia rhomboidalis*

SIZE/ID: SVL 57mm. Similar to *C. longipes* with 1–2 large pointed lobules on anterior ear. Brown above with bright blue chin and contrasting red throat.

HABITAT/RANGE: Rainforests and margins in tropical mid-eastern Qld.

BEHAVIOUR: Basks in sunlit patches.

Magnetic Island, Qld.

SCINCIDAE

Orange-flanked Rainbow Skink *Carlia rubigo*

Hallett State Forest, Qld.

SIZE/ID: SVL 44mm. Limbs well-developed; four fingers and five toes. Ear-opening with 1–2 large rounded anterior lobules. Dorsal scales hexagonal with three keels. ♂ has orange flanks and blue throat with dark speckling. ♀ has white midlateral stripe. Numerous similar species.

HABITAT/RANGE: Dry forests in eastern Qld.

BEHAVIOUR: Sun-loving. Basks in leaf litter.

Southern Rainbow Skink *Carlia tetradactyla*

SIZE/ID: SVL 64mm. Similar to *C. rubigo* with one large blunt anterior ear-lobule. Dorsal scales smooth. Bluish-green flanks with red upper and lower lateral stripes.

HABITAT/RANGE: Dry forests and woodlands with tussocks in south-east.

BEHAVIOUR: Often lays eggs in active ant nests.

Oakey, Qld.

SCINCIDAE

Three-toed Snake-tooth Skink
Coeranoscincus reticulatus

Adult and juvenile. Cunningham's Gap, Qld.

SIZE/ID: SVL 195mm. Very long-bodied, with widely-spaced, very short limbs with three digits, highly polished scales and rounded snout. Ears represented by scaly depressions. Adult brown to grey with little pattern. Juvenile cream to brown with prominent bands on anterior body and dark patches on eyes and ear-depressions.

HABITAT/RANGE: Subtropical rainforests between north-eastern NSW and Fraser Island, Qld.

BEHAVIOUR: Burrower. Dwells in and under damp rotting logs. Long recurved teeth believed to help when eating slippery earthworms. Egg-layer.

SCINCIDAE

Fraser Island Sand Skink *Coggeria naufragus*

Central Station, Fraser Island.

SIZE/ID: SVL 127mm. Very long-bodied, with widely spaced very short limbs with three digits, highly polished scales and wedge-shaped snout. Ears represented by scaly depressions. Pale brown with narrow dark longitudinal lines on back, a dark dorsolateral line of dark dots and grey flanks.

HABITAT/RANGE: Tall forests and heaths growing on sand. Endemic to Fraser Island, Qld.

BEHAVIOUR: Burrower. Dwells under damp rotting logs, and probably deeper into sandy substrate.

SCINCIDAE

Martin's Skink *Concinnia martini*

Sunshine Coast, Qld.

SIZE/ID: SVL 70mm. Well-developed limbs with five digits, smooth shiny scales, no ear-lobules and moveable scaly lower eyelids. Coppery-brown with black chequered pattern on back, broad ragged-edged black lateral zone and numerous narrow black bands on original tail. Several similar species in eastern Australia.

HABITAT/RANGE: Woodlands forests, outcrops and some urban areas along east coast and adjacent interior.

BEHAVIOUR: Diurnal, basking on logs and rocks and retreating to cavities and crevices. Live-bearer.

SCINCIDAE

Buchanan's Snake-eyed Skink
Cryptoblepharus buchananii

Wembley, WA.

SIZE/ID: SVL 49mm. Dorsally depressed. Long limbs with five long slender digits. Lower eyelid fused to form a fixed spectacle surrounded by granular scales. Greyish-brown with black dorsal flecks and broad ragged pale paravertebral stripes. Numerous similar species across Australia. Identification can be problematic.

HABITAT/RANGE: Variety of habitats including urban. Western WA.

BEHAVIOUR: Favours vertical surfaces, shinning swiftly over wood, rock, walls and fences. Diurnal. Egg-laying.

SCINCIDAE

Juno's Snake-eyed Skink *Cryptoblepharus juno*

Purnululu National Park, WA.

SIZE/ID: SVL 43mm. Extremely dorsally depressed. Very long limbs with five long slender digits. Lower eyelid fused to form a fixed spectacle surrounded by granular scales. Reddish-brown with small irregular dark flecks. Similar species on outcrops across northern Australia. Identification can be problematic.

HABITAT/RANGE: Rock outcrops, cliffs and gorges of eastern Kimberley region, WA, and adjacent NT.

BEHAVIOUR: Rock-inhabiting. Favours vertical surfaces, shinning swiftly rock faces. Diurnal. Egg-laying.

SCINCIDAE

Elegant Snake-eyed Skink *Cryptoblepharus pulcher*

Kurwongbah, Qld.

SIZE/ID: SVL 41mm. Dorsally depressed. Long limbs with five long slender digits. Lower eyelid fused to form a fixed spectacle surrounded by granular scales. Greyish-brown, broadly black-edged vertebral stripe and prominent straight-edged pale dorsolateral stripes. Two subspecies. Numerous similar species across Australia. Identification can be problematic.

HABITAT/RANGE: Variety of habitats including urban along eastern Australia between southern NSW and northern Qld, and across Great Australian Bight

BEHAVIOUR: Favours vertical surfaces, shinning swiftly over wood, rock, walls and fences. Diurnal. Egg-laying.

SCINCIDAE

Greer's Ctenotus *Ctenotus greeri*

Barkly region, NT.

SIZE/ID: SVL 65mm. Well-developed limbs with five digits. Anterior ear-lobules. Moveable scaly lower eyelid. Pattern of stripes and spots: pale-edged dark vertebral stripe, pale spots or dashes along outer back, white dorsolateral stripe, white midlateral stripe broken anteriorly, reddish flanks with white spots. Numerous similar species. Identification can be problematic.

HABITAT/RANGE: Central and western deserts, often with spinifex, on sand and loam.

BEHAVIOUR: Extremely swift, diurnal and terrestrial, dashing between low vegetation. Shelters in burrows. Egg-layer.

SCINCIDAE

Red-legged Ctenotus *Ctenotus labillardieri*

Peaceful Bay, WA.

SIZE/ID: SVL 75mm. Well-developed limbs with five digits. Anterior ear-lobules. Moveable scaly lower eyelid. Back plain brown in north; black-flecked in south. Continuous pale dorsolateral stripes, reddish limbs with black marbling, black flanks, white midlateral stripe and yellow belly. Other similar species.

HABITAT/RANGE: South-western WA, from high rainfall forest to heaths and rock outcrops.

BEHAVIOUR: Penetrates moist habitats more effectively than other *Ctenotus*, a group including many dry terrain and desert species.

SCINCIDAE

Leopard Ctenotus *Ctenotus pantherinus*

Ballera, Qld.

SIZE/ID: SVL 90–114mm. Well-developed limbs with five digits. Anterior ear-lobules. Moveable scaly lower eyelid. Olive to rich reddish-brown with 8–12 rows of prominent dark-edged longitudinal white dashes. Several subspecies across Australia.

HABITAT/RANGE: Dry to arid shrublands and spinifex grasslands across the centre, dry tropical north and interior of the south-west.

BEHAVIOUR: Diurnal and terrestrial, dashing between low vegetation, particularly spinifex. Shelters in burrows. Eats invertebrates and other vertebrates such as smaller skinks. Egg-layer.

SCINCIDAE

Robust Striped Ctenotus *Ctenotus robustus*

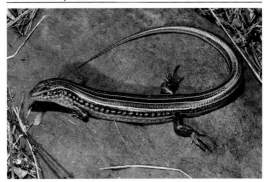

Warrill View, Qld.

SIZE/ID: SVL 123mm. Well-developed limbs with five digits. Anterior ear-lobules. Moveable scaly lower eyelid. Variable. Brown above with white-edged black vertebral and cream dorsolateral stripes, olive upper flanks with diffuse pale spots or blotches, pale midlateral stripe reaching forward under eye. Patternless populations on some Qld sand islands. Many similar species, with problematic identification.

HABITAT/RANGE: Widespread over east, centre and north.

BEHAVIOUR: Diurnal and terrestrial. One of the largest and most widespread of the striped skinks. Includes smaller skinks on its diet. Egg-layer.

SCINCIDAE

Stripe-headed Ctenotus *Ctenotus striaticeps*

Mt Isa area, Qld.

SIZE/ID: SVL 50mm. Well-developed limbs with five digits. Anterior ear-lobules. Moveable scaly lower eyelid. Black with 6–8 very sharp yellow stripes, suffused anteriorly with orange. These include an inner pair extending forward to converge on snout.

HABITAT/RANGE: Stony soils vegetated with spinifex in semi-arid north-western Qld and adjacent NT.

BEHAVIOUR: Diurnal and terrestrial. Secretive, seldom venturing far into open spaces, tending to forage close to edges of spinifex clumps. Egg-layer.

SCINCIDAE

Tasmanian She-oak Skink
Cyclodomorphus casuarinae

Ellendale, Tas.

SIZE/ID: SVL 174mm. Long-bodied and short-limbed with five digits. Long cylindrical tail. Parietal scales separated. Grey to rich reddish-brown, occasionally black. Usually dark sides to scales create irregular broken stripes, but sometimes pattern absent. Similar species on mainland.

HABITAT/RANGE: Endemic to Tas. Open grassy habitats and forest edges, typically with tussocks or heath.

BEHAVIOUR: Related to more familiar blue-tongues. If threatened gapes mouth, flickers blue tongue and may rear body like small venomous snake. Omnivorous, taking fruits and invertebrates. Live-bearer.

SCINCIDAE

Western Slender Blue-tongue
Cyclodomorphus celatus

City Beach, WA.

SIZE/ID: SVL 121mm. Long-bodied and short-limbed with five digits. Parietal scales separated. Grey to white with dark streaks on most body scales, with each streak extending full length of scale.

HABITAT/RANGE: Sandy and limestone-based coastal regions including sand flats and beach dunes along lower west coast. Common in dunes behind Perth's suburban beaches.

BEHAVIOUR: Shelters in thick low vegetation and under leaf litter. Omnivorous, taking insects, small snails and fruits. Live-bearer.

SCINCIDAE

Pink-tongued Skink *Cyclodomorphus gerrardii*

Mt Glorious, Qld.

SIZE/ID: SVL 200mm. Long-bodied and short-limbed with five digits. Long cylindrical, prehensile tail. Parietal scales separated. Tongue usually pink; blue on juvenile. Pattern ranges from prominent grey and black bands, to plain brown with dark snout. Juvenile pattern very sharp and obvious.

HABITAT/RANGE: Moist eastern areas between Blue Mountains, NSW, and north Qld.

BEHAVIOUR: Semi-arboreal, making good use of claws and prehensile tail. Diurnal and nocturnal. When harassed, gapes mouth and flickers tongue. Eats mainly molluscs, and has enlarged rounded rear teeth to crack snail shells.

SCINCIDAE

Cunningham's Skink *Egernia cunninghami*

Girraween National Park, Qld.

SIZE/ID: SVL 200mm. Robust. Limbs with five digits, long tail round in cross-section and large spine on each dorsal scale, longest on tail. Parietal scales separated. Colour varies geographically, including: plain or speckled brown; blackish with cream blotches; grey or reddish-brown with dark bands.

HABITAT/RANGE: Rock outcrops and woodlands from south-eastern Qld to eastern SA.

BEHAVIOUR: Shelters in rock or wood crevices, using spines to secure itself and hinder extraction. Communal, living in family groups and sharing a common latrine site. Omnivorous. Live-bearing.

SCINCIDAE

King's Skink *Egernia kingii*

Houtman Abrolhos, WA. [Photo credit. G. Harold]

SIZE/ID: SVL 244mm. Robust. Limbs with five digits, long tail round in cross-section and several low keels on each dorsal scale. Parietal scales separated. Dark grey to black. Juveniles and some adults weakly to intensely marked with pale spots.

HABITAT/RANGE: Open forest and heaths on south-western WA including many offshore islands.

BEHAVIOUR: Shelters in crevices and burrows. Omnivorous, often including seabird eggs in its diet. Live-bearing.

SCINCIDAE

Yakka Skink *Egernia rugosa*

Alton National Park, Qld.

SIZE/ID: SVL 240mm. Robust. Limbs with five digits, long thick tail, round in cross-section and several low keels on each dorsal scale. Parietal scales fragmented. Large plate-like ear-lobules. Brown with broad darker mid-dorsal zone.

HABITAT/RANGE: Woodlands and rock outcrops in eastern interior of Qld.

BEHAVIOUR: Occupies communal burrow systems under rocks and logs, including disused rabbit warrens. Uses shared latrine site Omnivorous, taking significant vegetation. Live-bearing.

SCINCIDAE

Black Rock Skink *Egernia saxatilis*

Warrumbungles National Park, NSW.

SIZE/ID: SVL 135mm. Robust and dorsally depressed. Limbs with five digits, and 2–5 sharp keels on each dorsal scale. Parietal scales separated. Usually dark grey to black; often dark brown with broken lines of black dashes in Warrumbungle Mts, NSW. Two subspecies.

HABITAT/RANGE: Rock outcrops, occasionally woodlands, in south-east. Warrumbungles population is outlying northern subspecies.

BEHAVIOUR: Shelters in rock crevices, occasionally cracks in logs. Communal, living in small colonies. Omnivorous. Live-bearing.

SCINCIDAE

Gidgee Skink *Egernia stokesii*

Broken Hill, NSW.

SIZE/ID: SVL 158–194mm. Very robust. Limbs with five digits, short flat tail, two large spines on each dorsal scale, and larger single spines on tail. Parietal scales separated. Colour varies geographically, including: reddish-brown to black with large prominent pale blotches in west; pale brown with variable small pale blotches in east. Several subspecies.

HABITAT/RANGE: Rocks, woodlands and shrublands in fragmented range across Australia.

BEHAVIOUR: Shelters in rock crevices and fallen logs, using spines to wedge themselves securely. Communal, living in small colonies. Omnivorous. Live-bearing.

SCINCIDAE

Mangrove Skink *Emoia atrocostata*

Boigu Island, Qld.

SIZE/ID: SVL 85mm. Well-developed limbs with five digits. Moveable lower eyelid enclosing transparent window. Frontoparietal scales fused into one shield. Brown with pale spots and dark flecks above, with broad black ragged-edged upper flanks.

HABITAT/RANGE: Extreme coastal rocks and mangroves on tip of Cape York and Torres Strait islands, Qld. Also Pacific islands and South-East Asia.

BEHAVIOUR: Basks on exposed sites including within splash zone, avoiding waves and moving in to forage for invertebrates as they recede. Egg-layer.

SCINCIDAE

Northern Narrow-banded Skink
Eremiascincus intermedius

Barkly region, NT.

SIZE/ID: SVL 88mm. Moderately well-developed limbs with five digits. Moveable scaly lower eyelid. Frontoparietal scales divided, and seven scales along upper lip. Smooth glossy body scales with low ridges on hips and tail. Reddish-brown with 6–16 oblique narrow bands across body and up to 42 neat transverse bands across original tail. Other similar desert species.

HABITAT/RANGE: Sandy deserts with spinifex in central NT and adjacent Qld.

BEHAVIOUR: Nocturnal and crepuscular. Able to dive into and move with ease through loose sand. Egg-layer.

SCINCIDAE

Northern Bar-lipped Skink *Eremiascincus isolepis*

Broome, WA.

SIZE/ID: SVL 72mm. Moderately well-developed limbs with five digits. Moveable scaly lower eyelid. Frontoparietal scales divided, and six scales along upper lip. Smooth glossy body scales. Brown with darker dorsal spots, becoming more concentrated along upper flanks. Similar species across north.

HABITAT/RANGE: Northern Australia, seeking moist shelter sites, from swamp margins and river flood plains to dry woodlands.

BEHAVIOUR: Nocturnal and crepuscular. Egg-layer.

SCINCIDAE

Elf Skink *Eroticoscincus graciloides*

Kurwongbah, Qld.

SIZE/ID: SVL 32mm. Tiny with large eyes, moveable lower eyelid enclosing a transparent disc, pointed snout, four fingers and five toes. Iridescent brown with weak dark dorsolateral line on body and two rusty stripes on tail.

HABITAT/RANGE: Shaded sheltered sites such as moist depressions within various wet and dry forests in south-east Qld.

BEHAVIOUR: Secretive, seldom venturing into open areas. Egg-layer.

SCINCIDAE

Brown Sheen Skink *Eugongylus rufescens*

Lockerbie Scrub, Qld.

SIZE/ID: SVL 169mm. Long body and tail, widely-spaced limbs with five digits, moveable scaly lower eyelid and parietal scales in contact. Brown with iridescent sheen. Juvenile has prominent pale bands.

HABITAT/RANGE: Monsoon forests on tip of Cape York, Torres Strait islands and New Guinea.

BEHAVIOUR: Shelters in and under large rotting logs and moist leaf litter. Egg-layer.

SCINCIDAE

Eastern Water Skink *Eulamprus quoyii*

Girraween National Park, Qld.

SIZE/ID: SVL 115mm. Robust with well-developed limbs with five digits, smooth shiny scales, moveable scaly lower eyelid and parietal scales in contact. Coppery-brown above with prominent yellow dorsolateral stripes and black upper flanks enclosing white spots. Other similar species.

HABITAT/RANGE: Waterside habitats such as river banks and swamp margins, but often extending into other sites away from water along eastern Australia.

BEHAVIOUR: Diurnal. Voracious predator of invertebrates, smaller lizards, frogs and tadpoles. May enter water to escape danger. Live-bearer.

SCINCIDAE

Black-tailed Bar-lipped Skink
Glaphyromorphus nigricaudis

Iron Range, Qld.

SIZE/ID: SVL 90mm. Robust with moderately well-developed limbs with five digits, smooth shiny scales, moveable scaly lower eyelid and parietal scales in contact. Brown with narrow dark bands across nape and fore-body, fading at midbody.

HABITAT/RANGE: Woodlands, monsoon forests and vine thickets of far north-eastern NT, Cape York and Torres Strait islands, Qld and southern New Guinea.

BEHAVIOUR: Nocturnal and crepuscular, sheltering under logs and leaf litter. Reported as live-bearing on southern Cape York; egg-laying elsewhere.

Fine-spotted Mulch Skink
Glaphyromorphus punctulatus

Charters Towers area, Qld.

SIZE/ID: SVL 70mm. Long-bodied with short, widely-spaced limbs with five digits, smooth shiny scales, moveable scaly lower eyelid and parietal scales in contact. Grey to brown with pattern, if present, comprising dark flecks or spots.

HABITAT/RANGE: Woodlands and vine thickets from mid-eastern to north-eastern Qld.

BEHAVIOUR: Nocturnal and crepuscular. Secretive, foraging mainly under cover and sheltering under logs, leaf litter and in soft soils and compost. Egg-layer.

SCINCIDAE

Prickly Forest Skink *Gnypetoscincus queenslandiae*

Millaa Millaa Falls, Qld.

SIZE/ID: SVL 84mm. Robust with moderately well-developed limbs with five digits and prominent raised keels on all scales creating a very rough texture. Dark brown with irregular narrow pale bands.

HABITAT/RANGE: Rainforest in the Wet Tropics, Qld.

BEHAVIOUR: Occupies shaded, damp sites in and under rotting logs and shuns direct sunlight. Extremely common is some areas with virtually all suitable sites harbouring a resident lizard. Slow-moving when disturbed. Feeds on invertebrates such as insects, gastropods and worms. Live-bearer.

SCINCIDAE

Beech Skink *Harrisoniascincus zia*

Lamington National Park, Qld.

SIZE/ID: SVL 55mm. Well-developed short limbs with five digits, smooth scales, moveable lower eyelid enclosing a transparent disc, frontoparietal scales paired and parietal scales in contact. Brown above with scattered pale and dark flecks, narrow pale dorsolateral stripe and dark grey upper flanks. Belly bright yellow.

HABITAT/RANGE: Montane rainforest and beech forest in south-eastern Qld and north-eastern NSW.

BEHAVIOUR: Basks in sunny patches, sheltering under logs and leaf litter. Egg-layer.

SCINCIDAE

Peron's Earless Skink *Hemiergis peronii*

Fowlers Bay, SA.

SIZE/ID: SVL 79mm. Long body and short limbs with three or four digits. Moveable lower eyelid with transparent disc. No ear-opening. Brown with 2–4 rows of dark dorsal dashes. Belly bright yellow. Two subspecies.

HABITAT/RANGE: Forests and heaths in southern SA and WA.

BEHAVIOUR: Shuns light. Shelters under damp logs and debris.

Eastern Earless Skink *Hemiergis talbingoensis*

Strathbogie Ranges, Vic.

SIZE/ID: SVL 60mm. Similar to *H. peronii*, with three digits. Brown with four rows of dark dorsal dashes. Belly yellow.

HABITAT/RANGE: Forests from Vic to southern NSW.

BEHAVIOUR: Similar to *H. peronii*.

SCINCIDAE

Murray's Skink *Karma murrayi*

Mary Cairncross Park, Qld.

SIZE/ID: SVL 108mm. Robust with well-developed limbs with five digits, smooth shiny scales, moveable scaly lower eyelid and parietal scales in contact. Coppery-brown above with dark flecks, dark lateral blotches near shoulder. Flanks mottled yellow and black, with dusting of fine bluish spots. Belly pale yellow.

HABITAT/RANGE: Subtropical rainforests of north-eastern NSW and south-eastern Qld.

BEHAVIOUR: Occupies cavities in fallen logs, often basking with head and fore-body protruding. Common and confiding near walking tracks. Eats invertebrates including funnel-web spiders. Live-bearing.

SCINCIDAE

Grass Skink *Lampropholis guichenoti*

Gerringong, NSW.

SIZE/ID: SVL 48mm. Small with well-developed limbs with five digits. Moveable lower eyelid enclosing transparent window. Frontoparietal scales fused into one shield. Brown to grey above, copper on head, with ragged-edged dark vertebral stripe, dark flanks and often a pale midlateral stripe. Other similar species.

HABITAT/RANGE: Most habitats including disturbed areas in southeast, between Kangaroo Island, SA and Cooloola, Qld.

BEHAVIOUR: Common garden lizard in towns and cities, basking beside paths, in woodheaps, etc. Generalist invertebrate feeder. Egg-layer, often depositing communally.

SCINCIDAE

Saxicoline Sun Skink *Lampropholis mirabilis*

Mt Elliot, Qld.

SIZE/ID: SVL 50mm. Small and dorsally depressed with long, well-developed limbs with five digits. Moveable lower eyelid enclosing transparent disc. Frontoparietal scales fused into one shield. Olive to copper with prominent dark blotches and white spots.

HABITAT/RANGE: Boulders and rocks associated with hoop pine and rainforest thickets on Magnetic Island and nearby mainland, Qld.

BEHAVIOUR: Rock-inhabiting. Agile and able to execute well-coordinated leaps between granite boulders. Egg-laying.

SCINCIDAE

Retro Slider *Lerista allanae*

SIZE/ID: SVL 88mm. Protrusive snout. Small ear-opening. Lower eyelid moveable. No fore-limbs, tiny hind-limb with one digit, rarely two. Grey with rows of black spots.

HABITAT/RANGE: Friable soils in Clermont/Capella area, Qld.

BEHAVIOUR: Burrower. Moves easily through soft soil. Egg-layer. Critically endangered.

Clermont area, Qld.

Limbless Fine-lined Slider *Lerista ameles*

SIZE/ID: SVL 58mm. Protrusive snout. Small ear-opening. Lower eyelid moveable. Limbs absent. Brown with six dark longitudinal lines.

HABITAT/RANGE: Soft soil in granite area near Mt Surprise, Qld.

BEHAVIOUR: Burrower, often in soil under rocks. Egg-layer.

Mt Surprise area, Qld.

SCINCIDAE

Western Two-toed Slider *Lerista bipes*

Barkly region, NT.

SIZE/ID: SVL 62mm. Protrusive snout. Small ear-opening. Lower eyelid moveable. No fore-limbs, small hind-limbs with two long digits. Yellowish to reddish-brown above with two narrow dark lines of dashes and broad black upper lateral stripe. Numerous similar species.

HABITAT/RANGE: Sandy deserts across WA and Central Australia.

BEHAVIOUR: Burrower, sheltering in sand, often under leaf litter. Streamlined snout and body allow easy movement through soft sand, leaving distinctive meandering trails. Egg-layer.

SCINCIDAE

Bougainville's Slider *Lerista bougainvillii*

Telowie Gorge, SA.

SIZE/ID: SVL 70mm. Weakly-protrusive snout. Small ear-opening. Lower eyelid moveable. Limbs present but very short, widely-spaced with five short digits. Grey to brown above, sometimes with four narrow dark dorsal lines, and broad black upper lateral stripe. Tail often flushed with yellow or orange, particularly on juvenile.

HABITAT/RANGE: Heaths and forests in south-eastern mainland and eastern Tas.

BEHAVIOUR: Shelters in loose soil under rocks and logs. Variable reproduction; live-bearing upland and egg-laying lowland populations.

SCINCIDAE

South Coast Five-toed Slider *Lerista microtis*

Peaceful Bay, WA.

SIZE/ID: SVL 52–60mm. Weakly-protrusive snout. Small ear-opening. Lower eyelid moveable. Limbs present but very short, widely-spaced with five short digits. Greyish-brown above with prominent stripes; broad black outer dorsal, white dorsolateral, black upper lateral and white lower lateral. Some populations also have two black dorsal stripes. Several subspecies.

HABITAT/RANGE: Heaths and coastal vegetation in south-west and disjunctly across Great Australian Bight, SA.

BEHAVIOUR: Shelters in dis-used ant nests and in loose soil under logs. Egg-layer.

SCINCIDAE

Inland Broad-striped Slider *Lerista nichollsi*

Lochada Nature Reserve, WA.

SIZE/ID: SVL 68mm. Protrusive snout. Small ear-opening. Lower eyelid fused to form a fixed spectacle. No fore-limbs, small hind-limbs with two digits. Very pale grey with broad dark vertebral stripe enclosing two rows of dark dots, broad black upper lateral stripe and yellow lower flanks and belly. Other similar species.

HABITAT/RANGE: Acacia shrublands in arid western interior of WA.

BEHAVIOUR: Burrower, sheltering in loose soil under leaf litter. Egg-layer.

SCINCIDAE

West Coast Worm Slider *Lerista praepedita*

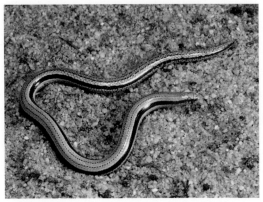

Moore River National Park, WA.

SIZE/ID: SVL 65mm. Slender body, flat protrusive snout. Small ear-opening. Lower eyelid moveable. No fore-limbs, hind-limbs reduced to minute stumps. Very pale greyish-brown above with two broken lines of dark dashes, dark upper lateral stripe and grey flush on tail.

HABITAT/RANGE: Heaths and woodlands along lower west coast.

BEHAVIOUR: Burrower moving easily through loose sand. Shelters in sand under leaf litter, rocks and logs. Egg-layer.

Outcrop Rock Skink *Liburnascincus mundivensis*

Amity Station, Qld.

SIZE/ID: SVL 56mm. Robust body, very long limbs and digits with four fingers and five toes, round ear-opening surrounded by pointed lobules, moveable lower eyelid enclosing transparent window and 2–3 keels on each dorsal scale. Olive with dark blotches, plain vertebral zone and pale flecks. Head copper.

HABITAT/RANGE: Outcrops and boulder scree from north-eastern to mid-eastern Qld and adjacent interior.

BEHAVIOUR: Rock-inhabiting. Alert and agile, foraging with ease over rock faces and making well-coordinated leaps between boulders. Egg-laying.

SCINCIDAE

Black Mountain Skink *Liburnascincus scirtetis*

Black Mountain, Qld.

SIZE/ID: SVL 64mm. Robust body, very long limbs and digits, four fingers and five toes, vertical ear-opening with long pointed lobules, protrusive eyes and upturned snout, moveable lower eyelid enclosing transparent window and weak keels on each dorsal scale. Adult black without pattern. Juvenile with greenish flecks forming weak stripes.

HABITAT/RANGE: Endemic to black boulders of Black Mountain near Cooktown, Qld.

BEHAVIOUR: Rock-inhabiting. Alert and agile, foraging with ease over rock faces and making well-coordinated leaps between boulders. Inquisitive but difficult to approach. Egg-laying.

SCINCIDAE

Desert Skink *Liopholis inornata*

Ethabuka, Qld.

SIZE/ID: SVL 84mm. Robust with deep blunt head, well-developed limbs with five digits, moveable scaly lower eyelid and parietal scales separated. Yellowish-brown to rich copper with rows of black flecks; longitudinal on back and vertical on flanks.

HABITAT/RANGE: Arid and semi-arid sandy areas including desert dunes, usually with spinifex, in central and southern Australia.

BEHAVIOUR: Digs burrows with one or more concealed exits at bases of shrubs or spinifex. Diurnal in mild weather; crepuscular to nocturnal in summer. Live-bearing.

SCINCIDAE

White's Skink *Liopholis whitii*

Port Albert, Vic.

SIZE/ID: SVL 113mm. Robust. Well-developed limbs with five digits, moveable scaly lower eyelid and parietal scales separated. Striped form has brown vertebral stripe, broad black dorsal stripes enclosing row of white spots, pale dorsolateral stripe and brown flanks with dark-edged white spots. Plain form has reduced back pattern.

HABITAT/RANGE: Forests and heaths, mainly where rocks are present, in south-eastern mainland and Tas.

BEHAVIOUR: Digs burrows, often under stones and logs. Often basks with head protruding. Communal. Live-bearer.

Girraween National Park, Qld.

SCINCIDAE

Swamp Skink *Lissolepis coventryi*

Tooradin, Vic.

SIZE/ID: SVL 100mm. Robust. Well-developed limbs with five digits, moveable scaly lower eyelid and parietal scales separated. Glossy black above with four coppery-bronze stripes. Flanks dark grey with pale spots.

HABITAT/RANGE: Margins of cool temperate wetlands including swamps and coastal salt marshes in southern Vic and far south-eastern SA.

BEHAVIOUR: Diurnal but secretive, basking among tussocks and other low vegetation and quickly fleeing to burrows if disturbed. Crustacean burrows often utilised. Live-bearer.

SCINCIDAE

Mourning Skink *Lissolepis luctuosa*

Lake Herdsman, WA.

SIZE/ID: SVL 129mm. Robust. Well-developed limbs with five digits, moveable scaly lower eyelid and parietal scales separated. Yellowish-brown above with 4–6 rows of angular black spots. Upper flanks black and lower flanks yellowish, marked with prominent yellow spots.

HABITAT/RANGE: Dense vegetation margining wetlands such as lakes, swamps, creeks and rivers in south-western WA.

BEHAVIOUR: Diurnal. Basks among tussocks and other thick vegetation, seldom venturing far from cover and quickly fleeing if disturbed. Live-bearer.

SCINCIDAE

Tree-base Litter Skink *Lygisaurus foliorum*

Kurwongbah, Qld.

SIZE/ID: SVL 39mm. Limbs well-developed; four fingers and five toes. Ear-opening horizontally elongate with one or more sharp to flat lobules. Lower eyelid fused to form a fixed spectacle. Dorsal scales smooth. Brown with fine pale peppering. Lips with dark flecks or bars. Breeding ♂ has orange throat, hind-limbs and tail.

HABITAT/RANGE: Dry forests and woodlands in eastern Australia between Townsville and Sydney.

BEHAVIOUR: Diurnal, basking and foraging among leaf litter at bases of trees and shrubs and, seldom venturing far from cover. Egg-layer.

SCINCIDAE

Sun-loving Litter Skink *Lygisaurus zuma*

Kirrima Range, Qld.

SIZE/ID: SVL 34mm. Limbs well-developed; four fingers and five toes. Ear-opening round to nearly horizontal with flat lobules around edges. Moveable lower eyelid enclosing transparent disc. Iridescent greyish-brown with copper head and dark flecks. Breeding. ♂ has red throat and tail. Other similar species.

HABITAT/RANGE: Tropical open forests and gully forests in mid-eastern Qld between Mackay and Kirrima.

BEHAVIOUR: Diurnal, foraging and basking among leaf litter, and seldom venturing far from cover. Egg-layer.

SCINCIDAE

Orange-speckled Forest Skink
Magmellia luteilateralis

Eungella National Park, Qld.

SIZE/ID: SVL 92mm. Robust with well-developed limbs with five digits, smooth shiny scales, moveable scaly lower eyelid and parietal scales in contact. Rich brown to bronze above, dark patch above fore-limb and burnt orange flanks enclosing prominent white spots.

HABITAT/RANGE: Montane tropical rainforests above 900m in Eungella National Park, Qld.

BEHAVIOUR: Diurnal. Occupies cavities in fallen logs, often basking with head and fore-body protruding. Live-bearer.

SCINCIDAE

Common Dwarf Skink *Menetia greyii*

Currawinya National Park, Qld.

SIZE/ID: SVL 38mm. Tiny. Limbs well-developed and widely-spaced with four fingers and five toes. Ear-opening minute without lobules. Lower eyelid fused to form a fixed spectacle. Dorsal scales smooth. Greyish-brown with dark dorsal dashes, dark upper flanks and white midlateral stripe. Among Australia's smallest reptiles. Other similar species.

HABITAT/RANGE: Widespread in dry habitats across Australia.

BEHAVIOUR: Diurnal, basking and foraging among leaf litter at bases of trees and shrubs and seldom venturing far from cover. Egg-layer.

SCINCIDAE

Shrubland Pale-flecked Morethia
Morethia obscura

Balga, WA.

SIZE/ID: SVL 56mm. Limbs well-developed with five digits. Frontoparietal scales fused to form a single shield. Lower eyelid fused to form a fixed spectacle. Brown to olive, sometimes with dark-edged white spots and weak pale midlateral stripe. Breeding ♂ has orange throat. Other similar species.

HABITAT/RANGE: Woodlands and shrublands across dry to semi-arid southern Australia.

BEHAVIOUR: Diurnal and terrestrial. Swift and alert, foraging and basking near and among leaf litter around low vegetation. Egg-layer.

SCINCIDAE

Fire-tailed Skink *Morethia ruficauda*

Ethabuka, Qld.

SIZE/ID: SVL 56mm. Limbs well-developed with five digits. Frontoparietal scales fused to form a single shield. Lower eyelid fused to form a fixed spectacle. Black with sharp white dorsolateral and midlateral stripes and fiery red tail. An additional white vertebral stripe in mid-western WA. Two subspecies.

HABITAT/RANGE: Dry to arid areas, often near rocks, in central, northern and north-western Australia.

BEHAVIOUR: Diurnal and terrestrial. Extremely swift, vanishing rapidly into leaf litter and vegetation if disturbed. Sinuously waves spectacular tail when foraging. Egg-layer.

SCINCIDAE

Nangur Spiny Skink *Nangura spinosa*

Nangur National Park, Qld.

SIZE/ID: SVL 95mm. Robust with well-developed limbs with five digits, strongly keeled to spiny scales, most pronounced on tail, and parietal scales separated. Brown with weak irregular dark dorsal bands, and yellowish bars on flanks.

HABITAT/RANGE: Restricted to vine thickets near Murgon and Kilkivan, south-eastern Qld.

BEHAVIOUR: Excavates burrows under roots, tree buttresses and rocks. Rests with head and forebody protruding at burrow entrance, seldom foraging. Ambush predator of small invertebrates. Live-bearer. Critically endangered. Threats include weed encroachment and potential illegal collecting.

SCINCIDAE

Ornate Snake-eyed Skink *Notoscincus ornatus*

Barkly region, NT.

SIZE/ID: SVL 39mm. Limbs well-developed with five digits. Frontoparietal scales fused to form one shield. Lower eyelid fused to form a fixed spectacle surrounded by granular scales. Brown to copper with 1–3 dorsal rows of black dots, upper lateral series of dark squarish bars or unbroken black upper lateral stripe, and pale midlateral stripe. Two subspecies.

HABITAT/RANGE: Sandy flats across tropical north and arid centre, extending into far northern woodlands and river margins.

BEHAVIOUR: Secretive sun-lover associated with leaf litter and low vegetation. Egg-layer.

SCINCIDAE

Lord Howe and Norfolk Islands Skink
Oligosoma lichenigera

Phillip Island via Norfolk Island.

SIZE/ID: SVL 80mm. Limbs well-developed with five digits. Frontoparietal scales paired. Moveable lower eyelid enclosing transparent disc. Brown to olive above with dark flecks, pale dorsolateral stripes and dark upper flanks enclosing pale spots and darker flecks.

HABITAT/RANGE: Largely restricted to rat-free islands off Lord Howe and Norfolk Islands. Formerly present on main islands.

BEHAVIOUR: Shelters under stones and low vegetation. Introduced rats have largely eliminated populations from main islands. Sun-loving but secretive. Egg-layers. Vulnerable species.

SCINCIDAE

Yolk-bellied Snake Skink *Ophioscincus ophioscincus*

Mt Glorious, Qld.

SIZE/ID: SVL 97mm. Completely limbless with 20–24 midbody scales, small eyes with moveable lower eyelid and no ear-opening. Silvery-white with 4–6 dorsal lines of dark dashes, black lateral stripe and yellow to orange belly. Similar species.

HABITAT/RANGE: Rainforests and moist sheltered areas within drier forests of south-eastern Qld.

BEHAVIOUR: Shelters in compost and in soil under rocks and logs. Secretive, avoiding sunlight. Feeds on invertebrates including earth worms. Egg-layer.

SCINCIDAE

Spinifex Snake-eyed Skink *Proablepharus reginae*

Barkly region, NT.

SIZE/ID: SVL 41mm. Slender limbs with five digits. Frontoparietal scales divided. Eyelid immoveable, a large fixed spectacle. Brown with dark edges to scales forming an obscure netted pattern. Breeding ♂ has red on neck and throat.

HABITAT/RANGE: Arid to semi-arid areas with spinifex on sand and loam across interior of WA and NT.

BEHAVIOUR: Extremely secretive, basking and foraging among leaf litter and spinifex and seldom venturing far from cover. Egg-layer.

SCINCIDAE

Tussock Skink *Pseudomoia pagenstecheri*

Falls Creek, Vic.

SIZE/ID: SVL 62mm. Limbs well-developed with five digits. Frontoparietal scales paired. Moveable lower eyelid enclosing large transparent disc. Supraciliary scales five. Brown with about five black dorsal stripes, and pale dorsolateral stripe on fouth scale row from midline. Pale midlateral stripe becomes orange to red on breeding ♂. Several very similar species.

HABITAT/RANGE: Tussock grasslands through Vic, eastern SA and eastern Tas to highlands of southern NSW.

BEHAVIOUR: Sun-loving, basking around tussocks. Live-bearer.

SCINCIDAE

Spencer's Skink *Pseudemoia spenceri*

Bogong, Vic.

SIZE/ID: SVL 65mm. Dorsally depressed head and body. Limbs long and well-developed with five digits. Frontoparietal scales paired. Moveable lower eyelid enclosing large transparent disc. Brown with broad black stripes along outer back, pale dorsolateral stripes, dark flanks and pale midlateral stripe.

HABITAT/RANGE: Cool temperate forests, woodlands and rock outcrops in Vic and south-eastern NSW.

BEHAVIOUR: Forages and basks on tree trunks and vertical rock surfaces, sheltering under bark and in crevices. Live-bearer.

SCINCIDAE

Magnetic Island Pygmy Skink
Pygmaeascincus sadlieri

Nelly Bay, Magnetic Island, Qld.

SIZE/ID: SVL 27mm. Tiny. Limbs well-developed and widely-spaced with four fingers and five toes. Ear-opening minute without lobules. Lower eyelid fused to form a fixed spectacle. Frontoparietal and interparietal scales fused to form one shield. Pale brown with contrasting darker flanks. One of Australia's smallest lizards.

HABITAT/RANGE: Dry forests, extending into disturbed areas such as home gardens on Magnetic Island, Qld.

BEHAVIOUR: Extremely secretive, basking and foraging among leaf litter and seldom venturing far from cover. Egg-layer.

SCINCIDAE

Three-toed Skink *Saiphos equalis*

Gerringong, NSW.

SIZE/ID: SVL 75mm. Long-bodied. Limbs short, widely-spaced with three digits. No ear-opening. Lower eyelid moveable and scaly. Shiny brown above with contrasting darker flanks. Ventral surfaces yellow to bright orange; black under tail.

HABITAT/RANGE: Variety of forests, woodlands, and grasslands in eastern Australia between southern Qld and southern NSW.

BEHAVIOUR: Shelters in compost and soft soil under rocks and logs. Unusual in having egg-laying populations in lowlands and live-bearing at higher elevations. Both modes uniquely recorded from one individual.

SCINCIDAE

Challenger Shade Skink *Saproscincus challengeri*

Lamington National Park, Qld.

SIZE/ID: SVL 57mm. Limbs well-developed with five digits, and long tail. Frontoparietal scales paired. Moveable lower eyelid enclosing transparent disc. Body scales not glossy. Brown, sometimes with scattered pale scales. Weak, often broken dark dorsolateral stripe on anterior body. Pale dorsal blotches on tail. Other similar species, and identification can be difficult.

HABITAT/RANGE: Subtropical rainforests in northern NSW and southern Qld.

BEHAVIOUR: Diurnal. Often active in shaded sites or dappled sunlight. Not particularly swift relative to other skinks. Egg-layer.

SCINCIDAE

Czechura's Shade Skink *Saproscincus czechurai*

Charmillin Ck, Qld.

SIZE/ID: SVL 34mm. Limbs well-developed with five digits, pointed snout and short tail. Frontoparietal scales paired. Moveable lower eyelid enclosing transparent disc. Body scales not glossy. Brown with dark and pale flecks and a rusty dorsolateral stripe. Breeding ♂ black with white spots on sides of head and neck, and orange on hind-limbs and tail.

HABITAT/RANGE: High altitude rainforests in Wet Tropics, Qld.

BEHAVIOUR: Diurnal but secretive. Lives among leaf litter, stones and logs, often near creeks. Egg-layer.

SCINCIDAE

Weasel Skink *Saproscincus mustelinus*

Lilydale, Vic.

SIZE/ID: SVL 55mm. Limbs well-developed with five digits, and long tail. Frontoparietal scales paired. Moveable lower eyelid enclosing transparent disc. Body scales not glossy. Brown to reddish-brown with pale dash behind eye and two prominent orange-brown stripes along tail.

HABITAT/RANGE: Cool temperate forests and heaths between southern Vic and New England Tablelands of northern NSW. Present in moist shaded parts of many suburban gardens.

BEHAVIOUR: Diurnal but seldom ventures far from cover. Egg-layer, recorded to sometimes lay communally.

SCINCIDAE

Bartle Frere Skink *Techmarscincus jigurru*

Mt Bartle Frere, Qld.

SIZE/ID: SVL 70mm. Dorsally depressed head and body. Limbs long and well-developed with five digits. Frontoparietal scales paired. Moveable lower eyelid enclosing large transparent disc. Copper above, with black dots, each with a pale rear edge, broken pale dorsolateral and midlateral stripes, and dark flanks with pale spots.

HABITAT/RANGE: Mist enshrouded boulder-scapes surrounded by dense rainforest above 1,400m altitude on Mt Bartle Frere, north Qld.

BEHAVIOUR: Diurnal. Adapted to cool conditions, basking on rocks and sheltering in crevices.

SCINCIDAE

Pygmy Blue-tongue *Tiliqua adelaidensis*

Burra, SA.

SIZE/ID: SVL 90mm. Long robust body, short widely-spaced limbs with five short digits, large head, short slender tail, scaly lower eyelid, and parietal scales separated. Brown to bluish grey, sometimes with black streaks or blotches.

HABITAT/RANGE: Treeless grasslands north of Adelaide, SA. Most or all known habitats are modified by grazing.

BEHAVIOUR: Shelters in vertical spider holes, rarely foraging and ambushing insects near burrow entrances. Once believed extinct, but rediscovered in 1992 after no records for 33 years. Live-bearer. Endangered.

SCINCIDAE

Southern or Blotched Blue-tongue
Tiliqua nigrolutea

Port Albert, Vic.

SIZE/ID: SVL 300mm. Very large and robust with short widely-spaced limbs with five short digits, thick tail, scaly lower eyelid, and parietal scales separated. Brown to black with large paler angular blotches. Darker with redder blotches in uplands of NSW.

HABITAT/RANGE: Temperate forests, woodlands and heaths between Tas and Blue Mountains, NSW.

BEHAVIOUR: Blue-tongues are among Australia's most popular lizards. Diurnal and slow-moving. Omnivorous, taking insects, birds' eggs, snails, fruits and soft foliage. Live-bearer.

SCINCIDAE

Shingleback, Bobtail or Sleepy Lizard *Tiliqua rugosa*

Gull Rock, WA.

SIZE/ID: SVL 260–310mm. Extremely distinctive. Very large and robust with wide triangular head, short limbs, large pinecone-like scales and short bulbous tail. Typically blackish-brown with paler markings, but bright splashes of orange often present in southern WA. Several subspecies.

HABITAT/RANGE: Dry to semi-arid shrublands, woodlands and pastoral areas across southern Australia.

BEHAVIOUR: Diet similar to blue-tongues. Mainly solitary but the same partners meet up each spring. If threatened, gapes pink mouth and distends broad bluish-purple tongue. Live-bearing, producing small litters of large offspring.

SCINCIDAE

Eastern and Northern Blue-tongue
Tiliqua scincoides

Miriam Vale, Qld.

SIZE/ID: SVL 300–320mm. Very large and robust with short widely-spaced limbs with five short digits, thick tail, scaly lower eyelid, and parietal scales separated. Variable but typically grey to black with prominent bands across back and tail. In north, bands are often fragmented on back and flushed with orange on flanks. Two subspecies.

HABITAT/RANGE: Widespread across east and north including within major capital cities.

BEHAVIOUR: Diet similar to other blue-tongues. Thrives in cities thanks to varied diet, sedentary habits of females and large litter sizes (up to 24).

DRAGONS Family Agamidae

More than 100 named species of dragons are found across Australia. Most live in deserts and tropics and just one reaches cool temperate Tasmania. Dragons share alert, upright postures, generally rough, non-glossy scales and long limbs and tails. Some are impressively adorned with spiny crests, erectable frills or beards and a number can change colour. Many breeding males acquire spectacular hues. They rely strongly on eyesight to locate prey and for territorial and courting displays, and include a repertoire of communication signals including head bobs and dips, tail-lashes, arm-waves and even colour flashes in the ultra violet spectrum.

A Slater's Ring-tailed Dragon *(Ctenophorus slateri)* surveys its surroundings from atop a piece of roadside debris. Mt Isa, Qld.

AGAMIDAE

Jacky Dragon *Amphibolurus muricatus*

Girraween National Park, Qld.

SIZE/ID: SVL 120mm. Irregularly-sized scales on back, including five enlarged longitudinal rows. Thigh covered with irregular mix of large and small scales. Grey with two broad pale dorsal stripes, often deeply notched or broken into a series of blotches. Mouth-lining yellow.

HABITAT/RANGE: Dry open forests, woodlands and heaths from south-eastern SA to south-eastern Qld.

BEHAVIOUR: Perches on stumps, trunks and branches. If pursued it can raise its body and sprint on the hind legs. ♂ engage in complex displays including head bobbing and tail lashing. The temperature of egg-incubation determines the sex of the offspring.

AGAMIDAE

Chameleon Dragon *Chelosania bruneae*

Dampier Peninsula, WA.

SIZE/ID: SVL 118mm. Laterally compressed head and body, relatively short limbs, blunt-tipped tail and small eye-opening surrounded by granular scales. Grey to brown with dark lines radiating from eye, dark variegations on body and banded tail.

HABITAT/RANGE: Tropical woodlands across northern Australia between Dampier Peninsula, WA and far north-western Qld.

BEHAVIOUR: Arboreal and slow-moving. Extremely cryptic and normally encountered on ground crossing roads. If harassed, gapes mouth and expands throat.

AGAMIDAE

Frill-neck *Chlamydosaurus kingii*

El Sharana, NT.

SIZE/ID: SVL 258mm. One of Australia's largest dragons. A large frill lies folded around neck and shoulders – standing out erect to encircle head. Grey to brown with frill normally grey in Qld; splashed with red in NT and WA.

HABITAT/RANGE: Tropical woodlands and dry forests across north, from to Broome, WA to Brisbane, Qld.

BEHAVIOUR: Arboreal, perching vertically on rough-barked trees. Descends to ground feed on insects. The frill is erected, with mouth agape, to defend against predators and for displays between rival ♂.

AGAMIDAE

Gravel Dragon *Cryptagama aurita*

Kimberley region, WA. [Photo credit: S. Mahony]

SIZE/ID: SVL 48mm. Round head, rotund body, obvious ear-opening, short limbs, and tail shorter than head and body. Dorsal scales irregular with small, granular scales mixed with scattered rounded enlarged tubercles. A fringe of scales along upper lip. Reddish-brown to brick red.

HABITAT/RANGE: Dry stony areas with spinifex in Kimberley region, WA and adjacent NT.

BEHAVIOUR: Very poorly known. Secretive and closely resembling the pebbles among which it rests.

AGAMIDAE

Western Heath Dragon *Ctenophorus adelaidensis*

Melaleuca Park, WA.

SIZE/ID: SVL 52mm. Small with short limbs and tail. Dorsal scales irregular with small scales and enlarged raised tubercles, mainly on areas of dark colour. Enlarged spiny scales along sides of tail-base, and an angular ridge along outer lower edge of jaw. Grey with dorsal and upper lateral rows of angular dark blotches. Black (♂) or grey (♀) stripes on belly.

HABITAT/RANGE: Lower west coast of WA, in Banksia woodlands and heath on pale sand.

BEHAVIOUR: Active on open sand along edges of vegetation. Does not use raised objects as perching sites.

AGAMIDAE

Crested Dragon *Ctenophorus cristatus*

Lake Cronin, WA.

SIZE/ID: SVL 110mm. Long limbs and tail. Dorsal scales irregular, mostly small and uniform with enlarged spiny vertebral crest, and spiny scales along each shoulder. Cream, yellow to bright orange with black network on fore-body, plain grey on hind body, and cream to orange with black rings on tail.

HABITAT/RANGE: Dry to semi-arid woodlands from southern interior of WA to Eyre Peninsula, SA.

BEHAVIOUR: Active on the ground and perches on low timber. Extremely swift, running on hind-limbs, leading to alternative name of 'bicycle lizard'.

AGAMIDAE

Tawny Dragon *Ctenophorus decresii*

Male. Telowie Gorge, SA.

SIZE/ID: SVL 82mm. Head and body dorsally depressed. Body scales uniform. Sexes differ. ♂ bluish grey with black flanks (north) or blotches (south), with white, yellow to red along upper edge. Throat blue (south) or orange, yellow to grey (north). ♀ brown with dark flanks and scattered dark flecks.

HABITAT/RANGE: Rock-inhabiting, between Kangaroo Island, Flinders Ranges and Olary Spur, SA.

BEHAVIOUR: Perches on elevated rocks. Flattened body allows access to narrow crevices.

Males display with raised bodies and horizontally coiled tails.

Female. Telowie Gorge, SA.

AGAMIDAE

Lake Eyre Dragon *Ctenophorus maculosus*

Lake Eyre, SA.

SIZE/ID: SVL 69mm. Robust body, round head and relatively short limbs and tail. Body scales uniform. Ear-opening completely hidden. Whitish with fine dark flecks, two dorsal rows of circular black blotches and a black streak on chin. Red ventral flush when breeding.

HABITAT/RANGE: Featureless salt lakes in arid SA.

BEHAVIOUR: Burrows under salt crusts. Perches on driftwood and salt crusts. Dominant ♂ defend the best sites. ♀ flip on backs to deter unwanted mating. During rare floods, entire populations forced onto surrounding sand hills.

AGAMIDAE

Central Netted Dragon *Ctenophorus nuchalis*

Hay River, NT.

SIZE/ID: SVL 115mm. Robust body, round head and relatively short limbs and tail. Body scales uniform. Yellowish-brown with dark netted pattern and pale vertebral stripe. Heads of mature ♂ reddish. Femoral pores curve forward to anterior thigh.

HABITAT/RANGE: Vast semi-arid to arid areas from west coast to interior of NSW.

BEHAVIOUR: Perches atop rocks, timber and road-side debris, retreating to one of several burrows to escape heat and predators. Burrows are sealed during winter inactivity. One of the most conspicuous desert lizards.

AGAMIDAE

Ornate Dragon *Ctenophorus ornatus*

Charles Darwin Nature Reserve, WA.

SIZE/ID: SVL 93mm. Head and body extremely dorsally depressed. Body scales uniform. ♂ are reddish-brown to black with pale dorsal blotches. Some inland populations redder, and blotches replaced by vertebral stripe. Tail black and white ringed. ♀ have weaker pattern with faint bands on tail.

HABITAT/RANGE: Granite sheets and outcrops in south-western WA.

BEHAVIOUR: Perches on rocks and sprints rapidly on all fours across open granite expanses. Flat body allows access to narrow crevices.

AGAMIDAE

Painted Dragon *Ctenophorus pictus*

Ballera area, Qld.

SIZE/ID: SVL 65mm. Robust body, round head and relatively short limbs and tail. Body scales uniform. ♂ are variable; brown to orange with bluish vertebral stripe and dark bars. When breeding, head flushed with red, orange to yellow and blue over throat and limbs. ♀ duller.

HABITAT/RANGE: Semi-arid south coast and interior, on sand with shrubs, spinifex or canegrass.

BEHAVIOUR: Excavates U-shaped burrow with concealed exit at base of shrub. Mainly terrestrial, utilising low perches. Brightly coloured ♂ display with raised crest and vertebral ridge.

AGAMIDAE

Eastern Mallee Sand Dragon
Ctenophorus aff. spinodomus

Hattah-Kulkyne National Park, Vic.

SIZE/ID: SVL 48mm. Very long limbs and tail. Body scales uniform. Brown with prominent pale stripes along back and sides edged above between and below with black blotches. ♂ have black chests with spotted throats. Similar species.

HABITAT/RANGE: Semi-arid mallee with spinifex in north-eastern Vic and adjacent SA, south of Murray River.

BEHAVIOUR: Completely terrestrial, never using elevated perches. Extremely swift, dashing across open sand between spinifex clumps. Populations are annual, with most lizards dying each year.

AGAMIDAE

Red-barred Dragon *Ctenophorus vadnappa*

Blinman Creek, SA.

SIZE/ID: SVL 85mm. Moderately depressed head and body. Body scales uniform. Breeding ♂ are spectacular, with red bars on black flanks, blue vertebral stripe, and throat striped with blue and yellow. ♀ are dull brown with dark flecks.

HABITAT/RANGE: Dry rocky areas in Flinders Ranges and on outcrops north of Lake Torrens, SA.

BEHAVIOUR: Rock-inhabiting, perching prominently on raised vantage points. Rival males perform impressive displays, laterally compressing and raising their bodies with tails coiled vertically over their backs.

AGAMIDAE

Mulga Dragon *Diporiphora amphiboluroides*

Mt Clere Station, WA.

SIZE/ID: SVL 94mm. Short limbs, upturned snout and blunt-tipped tail. Dorsal scales irregular; small mixed with five rows of enlarged scales including crest on nape. Grey with longitudinal dark streaks creating a cryptic bark-like pattern.

HABITAT/RANGE: Arid to semi-arid mulga shrublands and woodlands in southern interior of WA.

BEHAVIOUR: Arboreal though sometimes found in leaf litter or crossing roads. Secretive and slow-moving, blending artfully with its woody backdrop. When encountered, slides slowly from view rather than dashing to safety.

AGAMIDAE

Tommy Round-head *Diporiphora australis*

Karawatha State Forest, Qld.

SIZE/ID: SVL 65mm. Long limbs and tail. Gular, scapular and postauricular folds present. Body scales uniform but with ridges along midline and outer back. Grey to reddish-brown, with varying combinations of two narrow pale stripes overlying broad dark bars. Sometimes patternless.

HABITAT/RANGE: Woodlands and dry forests in eastern Qld.

BEHAVIOUR: Perches on rocks and low woody debris.

AGAMIDAE

Nobbi Dragon *Diporiphora nobbi*

Girraween National Park, Qld.

SIZE/ID: SVL 72mm. Long limbs and tail. Gular, scapular and postauricular folds present. Body scales irregular, including five rows of enlarged scales. Two narrow yellow stripes overly about six broad dark bars. Tail-base of breeding ♂ often flushed with pink to mauve.

HABITAT/RANGE: Woodlands, dry forests and mallee from northern Qld to eastern SA.

BEHAVIOUR: Perches on rocks and low woody debris. Common though believed sensitive to habitat disturbance.

AGAMIDAE

Superb Dragon *Diporiphora superba*

Surveyor's Pool, WA.

SIZE/ID: SVL 93mm. Extremely long slender body, limbs and tail. Body scales uniform. No gular, scapular or postauricular folds. Bright green, sometimes with vertebral reddish flush or occasionally a sharp, white-edged stripe.

HABITAT/RANGE: Sandstone areas of far north-western Kimberley, WA, mainly along vegetated edges of water courses.

BEHAVIOUR: Arboreal, dwelling among stems and thick foliage of shrubs. Slow-moving, relying on superb camouflage. Unique among mainland Australian lizards in being green.

AGAMIDAE

Long-nosed Dragon *Gowidon longirostris*

Barkly Tableland, NT.

SIZE/ID: SVL 93mm. Long body, long slender limbs and tail and long snout. Body scales uniform, with keels converging back towards midline. Rich reddish-brown with two prominent dorsal pale stripes and white stripe along lower jaw.

HABITAT/RANGE: Arid western and central Australia, along gorges and tree-lined water courses. Away from drainage lines they occupy scattered trees.

BEHAVIOUR: Arboreal but often descends to ground. Extremely swift, sprinting on hind-limbs to nearest tree if disturbed. Rival males display with raised nuchal crests.

AGAMIDAE

Water Dragon *Intellagama lesueurii*

Brisbane, Qld.

SIZE/ID: SVL 245mm. One of Australia's largest dragons. Robust with long powerful limbs and laterally compressed tail. Prominent spiny nuchal and vertebral crests. Eastern race: yellowish-brown with dark stripe behind eye. Breeding ♂ has red flush on chest. Gippsland race: breeding ♂ has throat black with yellow blotches and blackish chest. Two subspecies.

HABITAT/RANGE: Waterside habitats throughout eastern Australia.

BEHAVIOUR: Semi-arboreal and semi-aquatic, leaping from trees into water and swimming with ease. Can remain submerged for many minutes.

Genoa Falls, Vic.

AGAMIDAE

Horner's Lashtail *Lophognathus horneri*

King Edward River, WA.

SIZE/ID: SVL 86mm. Robust with long limbs and tail. Body scales have keels parallel to midline and include several slightly enlarged rows. Dark patch with white spot in ear. Breeding ♂ is black with two white dorsal stripes, discontinuous with broad white stripe through lips. ♀ is grey with weaker pattern. Similar species.

HABITAT/RANGE: Tropical woodlands and water courses across northern Australia.

BEHAVIOUR: Semi-arboreal, also foraging on ground. When disturbed sprints on hind-limbs. Arm-waving display has led to popular name, 'Ta-ta Lizard'.

AGAMIDAE

Boyd's Forest Dragon *Lophosaurus boydii*

SIZE/ID: SVL 150mm. Very laterally compressed body, long limbs and tail and angular brow. High spiny nuchal and vertebral crests disrupted above shoulders, plate-like scales on cheeks and line of spines along throat.

HABITAT/RANGE: Rainforests of Wet Tropics, Qld.

BEHAVIOUR: Slow-moving and cryptic, clinging to vertical stems and trunks.

Mossman Gorge, Qld.

Southern Angle-headed Dragon *Lophosaurus spinipes*

SIZE/ID: SVL 110mm. Similar to *L. boydii* but no plate-like scales on cheeks or line of spines along throat. Spiny nuchal and vertebral crests are continuous.

HABITAT/RANGE: Subtropical rainforests of Qld and NSW.

BEHAVIOUR: Similar to *L. boydii*.

Mt Glorious, Qld.

AGAMIDAE

Thorny Devil *Moloch horridus*

Alice Springs area, NT.

SIZE/ID: SVL 110mm. Robust body, short limbs and short, thick tail. Large bulb with two horn-like spines on neck. Body and appendages covered with large thorn-like spines, largest over each eye. Rich orange to olive blotches and pale vertebral stripe.

HABITAT/RANGE: Sandy arid to semi-arid areas from western Qld to west coast.

BEHAVIOUR: Slow-moving, walking like a clockwork toy. Feeds only on small black ants, with hundreds captured per meal. If harassed, lowers head, presenting the horned bulb on its neck as a decoy.

AGAMIDAE

Eastern Bearded Dragon *Pogona barbata*

Threat display. Brisbane, Qld.

SIZE/ID: SVL 266mm. Dorsally depressed. Body scales irregular: enlarged spines scattered over back, rows of slender spines along flanks, a curved row across base of head and bands of enlarged scales around tail. Spiny erectable gular pouch ('beard'). Grey, with two dorsal rows of pale blotches, varying according to mood and temperature.

HABITAT/RANGE: Open forests, woodlands, farmlands, parks and suburbs in eastern Australia.

BEHAVIOUR: Semi-arboreal, perching on stumps and fence posts. If harassed, flattens body, erects squarish spiny beard and gapes bright yellow mouth. Omnivorous.

AGAMIDAE

Downs Bearded Dragon *Pogona henrylawsoni*

Richmond area, Qld.

SIZE/ID: SVL 148mm. Dorsally depressed with short limbs and tail. Body scales irregular: enlarged spines scattered over back, rows of slender spines along flanks and a straight row across base of head. Erectable gular pouch ('beard') with few if any spines. Grey above with two rows of pale blotches.

HABITAT/RANGE: Treeless plains of cracking clay and Mitchell grass in semi-arid central Qld.

BEHAVIOUR: Perches on raised stones, tussocks or lumps of clay. Gapes mouth if harassed but beard is poorly developed.

AGAMIDAE

Central Bearded Dragon *Pogona vitticeps*

Blackall, Qld.

SIZE/ID: SVL 250mm. Dorsally depressed. Body scales irregular: enlarged spines scattered over back, rows of slender spines along flanks, and a straight row across base of head. Spiny erectable gular pouch ('beard'). Grey to brick red, with two dorsal rows of pale blotches, and sometimes with black beard. Colour varies according to mood and temperature.

HABITAT/RANGE: Semi-arid woodlands and shrublands in eastern interior of Australia.

BEHAVIOUR: Similar to Eastern Bearded Dragon but beard is more round and mouth is pink.

AGAMIDAE

Mountain Dragon *Rankinia diemensis*

Kinglake area, Vic.

SIZE/ID: SVL 82mm. Robust body, short limbs and tail. Body scales irregular: enlarged spines in four rows on back, a row along sides, scattered over flanks, and a row on each side of tail-base. Grey to reddish-brown with two broad pale dorsal stripes with zigzagging inner edges.

HABITAT/RANGE: Fragmented range; heaths and open forests including mountains in south-eastern mainland and Tas. Australia's most southerly dragon.

BEHAVIOUR: Terrestrial, seldom using elevated perches. Alert but not swift relative to other dragons.

AGAMIDAE

Swamplands Lashtail *Tropicagama temporalis*

SIZE/ID: SVL 120mm. Very long limbs and tail. Body scales uniform, with keels converging back towards midline. Grey to black with two very prominent white stripes continuous with broad stripe through lips.

HABITAT/RANGE: Woodlands, swamplands and river edges in coastal areas of far northern NT and Cape York, through Torres Strait islands to New Guinea.

BEHAVIOUR: Semi-arboreal, perching on trunks and limbs. Extremely swift, sprinting on hind-limbs when disturbed. Signals to other dragons with arm-waves and head-bobs.

Saibai Island, Qld.

AGAMIDAE

Smooth-snouted Earless Dragon
Tympanocryptis intima

Birdsville area, Qld.

SIZE/ID: SVL 61mm. Robust body, short limbs and tail. Body scales irregular, with about four irregular rows of enlarged spiny scales. Ear completely covered by scales. Grey to rich reddish-brown with two rows of about four large dark blotches.

HABITAT/RANGE: Stony deserts of Australia's eastern interior, often occupying very harsh and featureless terrain.

BEHAVIOUR: Terrestrial, selecting elevated perches on stones, sometimes during searing heat. Shelters in burrows or cavities under rocks.

AGAMIDAE

Grassland Earless Dragon *Tympanocryptis lineata*

Jerrabomberra, ACT.

SIZE/ID: SVL 55mm. Robust body, relatively short limbs and tail. Body scales irregular, with scattered enlarged spiny scales. Ear completely covered by scales. Grey to reddish-brown with three pale stripes overlying a series of short dark bars. Throats of breeding ♂ yellow marbled with black.

HABITAT/RANGE: Highly fragmented grasslands in ACT. One of several endangered earless dragon species in fragmented south-eastern grasslands.

BEHAVIOUR: Terrestrial, sheltering under rocks and in vertical invertebrate burrows. Critically endangered through habitat loss.

AGAMIDAE

Goldfields Pebble-mimic Dragon
Tympanocryptis pseudosephos

Wilthorpe Station, WA.

SIZE/ID: SVL 56mm. Rotund body, bulbous head, short limbs and tail. Body scales irregular, with short raised transverse rows of enlarged spiny scales. Ear completely covered by scales. Rich reddish-brown, sometimes with charcoal wash over back. Several similar species occur further north in WA.

HABITAT/RANGE: Harsh stony plains in arid interior of WA, south of Pilbara region.

BEHAVIOUR: An extraordinary pebble-mimic, crouching among stones with limbs tucked close. Virtually invisible unless it moves.

MONITORS or GOANNAS Family Varanidae

Goannas include Australia's largest lizards, which sometimes exceed 1.5m in length. There are also pygmy monitors less than 50cm long. Of 30 Australian species, most occupy deserts and tropical areas. Just one reaches the south-eastern mainland and there are none in Tasmania. Monitors have loose skin, bead-like non-glossy scales, powerful limbs, strong claws, a long snout and sharp, backward-curved teeth. They have keen eyesight, and their constantly probing, deeply forked tongues are acutely tuned chemical receptors for food, mates and danger. Egg-layers.

The Lace Monitor *(Varanus varius)* is a large, widespread large species found from south Vic to north Qld. Mt Mee, Qld.

VARANIDAE

Spiny-tailed Monitor *Varanus acanthurus*

Lawn Hill National Park, Qld.

SIZE/ID: TL 63cm. Robust with thick tail featuring bands of raised spines. Dark reddish-brown with pale stripes on neck, and dark-centred cream to yellow spots on back. Pattern more banded on islands off north-eastern NT. Two subspecies.

HABITAT/RANGE: Rock outcrops and areas with heavy soils featuring large termite mounds in tropical to semi-arid central and north-western Australia.

BEHAVIOUR: Shelters in rock crevices and cavities in termite mounds, using spiny tail to block entrance. Common in Australian husbandry.

VARANIDAE

Short-tailed Pygmy Monitor *Varanus brevicauda*

Central Australia, NT.

SIZE/ID: TL 23cm. Elongate body, short, widely-spaced limbs and short thick tail with strongly keeled scales. Reddish brown with scattered light and dark flecks that tend to form ocelli. World's smallest monitor. One similar species.

HABITAT/RANGE: Sandy spinifex deserts across WA and central Australia excluding Dampier Peninsula (replaced by the similar *V. sparnus*).

BEHAVIOUR: Extremely secretive, seldom venturing into open areas and tending to remain close to large spinifex hummocks. Feeds on insects and small lizards.

VARANIDAE

Desert Pygmy Monitor *Varanus eremius*

Barkly region, NT.

SIZE/ID: TL 46cm. Tail triangular in cross-section. Brown to rich red with pale and dark spots on back and prominent stripes along tail.

HABITAT/RANGE: Sandy spinifex deserts across WA and central Australia.

BEHAVIOUR: Shelters in burrows, usually at the base of or under spinifex hummocks. Extremely secretive and infrequently seen, through numerous tracks indicate relative abundance. Research in WA shows they readily investigate any fresh diggings.

VARANIDAE

Perentie *Varanus giganteus*

North West Cape, WA.

SIZE/ID: TL 240cm. Long neck, angular brow, powerful limbs and long, laterally compressed tail. Cream with dense dark speckling and large circular pale spots. Throat has prominent black net-like pattern. Largest Australian lizard.

HABITAT/RANGE: Arid zones from inland Qld to mid-west coast. Often associated with rocky hills and gorges but also occupies sandy deserts.

BEHAVIOUR: Forages widely over large home ranges, sheltering in caves, over-hangs and deep burrows. Diet includes mammals, reptiles including other monitors and venomous snakes, insects, eggs and carrion.

VARANIDAE

Pygmy Mulga Monitor *Varanus gilleni*

Ethabuka, Qld.

SIZE/ID: TL 38cm. Tail round in cross-section. Grey above, often with reddish flush, with narrow dark bands across body and base of tail, and longitudinal stripes along remainder of tail. Other similar species.

HABITAT/RANGE: Central and western arid zones featuring mulga, desert oak and other trees with loose bark or hollows.

BEHAVIOUR: Arboreal, sheltering behind loose bark and in hollow limbs. Feeds on invertebrates and small lizards such as geckos. Secretive, and though common is seldom observed active.

VARANIDAE

Gould's or Sand Monitor *Varanus gouldii*

Nappa Merrie Station, Qld.

SIZE/ID: TL 120–160cm. Tail laterally compressed. Typically yellow to brown or black with dark speckling, pale spots often dark centred and clustered, pale-edged dark stripe through eye, unmarked yellow tail-tip and grey V-shaped streaks on throat.

HABITAT/RANGE: Very widespread across sandy habitats in all mainland states. Often in open habitats such as woodlands or spinifex deserts.

BEHAVIOUR: Shelters in deep sloping burrows, with tracks and burrows often indicating presence in an area. Often digs for food such as burrowing lizards and reptile eggs.

VARANIDAE

Merten's Water Monitor *Varanus mertensi*

El Questro Station, WA.

SIZE/ID: TL 110cm. Tail strongly laterally compressed. Nostrils directed upwards on top of snout. Dark olive grey with small dark-edged pale spots.

HABITAT/RANGE: Tropical northern watercourses, typically lined with paperbarks, pandanus and other riverine vegetation, between Cape York, Qld and the Kimberley region, WA.

HABITAT/RANGE: Basks and forages among waterside rocks and vegetation. A superb swimmer, frequently entering water to elude danger and sometimes dropping from a considerable height. Also enters water to hunt for aquatic prey and can remain submerged for extended periods.

Yellow-spotted Monitor *Varanus panoptes*

Gascoyne Junction, WA.

SIZE/ID: TL 140cm. Tail laterally compressed. Typically reddish-brown or black with circular pale spots forming transverse rows, sometimes alternating with rows of larger black spots. Pale-edged dark stripe through eye, dark-banded yellow tail-tip and black spots on throat. Two subspecies.

HABITAT/RANGE: Northern Australia and the Pilbara and Gascoyne regions, WA. Often on hard soils and clays in woodlands and floodplains.

BEHAVIOUR: Terrestrial. Shelters in deep sloping burrows and forages extensively over large home ranges, often rearing on hind legs to survey its surroundings.

VARANIDAE

Spotted Tree Monitor *Varanus scalaris*

Burdekin River, Qld.

SIZE/ID: TL 60cm. Tail round in cross-section. Scales across top of head grade evenly into smaller scales above eyes. Extremely variable. Typically grey to black with bands of large, dark-centred pale spots. These may be fragmented or sometimes overlaid with reddish brown bands. Probably a species complex. Similar species.

HABITAT/RANGE: Timbered areas, from open woodlands to rainforests, across northern Australia.

BEHAVIOUR: Arboreal, sheltering in hollow limbs and trunks.

VARANIDAE

Black-headed or Freckled Monitor *Varanus tristis*

Morven, Qld.

SIZE/ID: TL 76cm. Tail round in cross-section. Scales across top of head sharply delineated from smaller scales above eyes. Grey to black with numerous dark centred cream to reddish-brown ocelli arranged in transverse bands. Sometimes head, neck and tail black without pattern. Two subspecies. Similar species.

HABITAT/RANGE: Dry to arid timbered and rocky areas, across northern, central and western Australia.

BEHAVIOUR: Arboreal and rock-inhabiting, sheltering under bark, in hollows and in rock crevices. Frequently utilises human structures.

Lace Monitor *Varanus varius*

Conondale Ranges, Qld.

SIZE/ID: TL 210cm. Tail laterally compressed. Grey to black with cream spots, sometimes forming bands. Black and yellow bands across chin, and very broad dark and pale bands on tail. Banded ('Bell's') phase has broad, simple yellow and brown to black bands across body. Juveniles of both forms intensely patterned.

HABITAT/RANGE: Timbered habitats of eastern Australia between Wet Tropics and south-eastern SA.

BEHAVIOUR: Arboreal, often ascending large trees, and forages extensively on ground. Nests in active termite mounds; termites heal the scar to seal in the eggs.

FURTHER READING

Bush, B., Maryan, B., Browne-Cooper, R. and Robinson, B. 2007. *Reptiles and Frogs in the Bush: Southwestern Australia*. University of Western Australia.

Cogger, H. 2014. *Reptiles and Amphibians of Australia*. Sixth Edition. CSIRO Publishing.

Hutchinson, M., Swain, R. and Driessen, M. 2001. *Snakes and Lizards of Tasmania*. University of Tasmania.

Melville, J. and Wilson, S., 2019. *Dragon Lizards of Australia. Evolution, Ecology and a Comprehensive Field Guide*. Museums Victoria Publishing.

Robertson, P. and Coventry, A. J., 2019. *Reptiles of Victoria*. CSIRO Publishing.

Swan, M. (Ed), 2008. *Keeping and Breeding Australian Lizards*. Mike Swan Herp Books.

Swan, M. and Watherow, S., 2005. *Snakes, Lizards and Frogs of the Victorian Mallee*. CSIRO Publishing.

Swan, G., Sadlier, R. and Shea, G., 2017. *A Field Guide to Reptiles of New South Wales*. Third Edition. Reed New Holland.

Williams, C. and Maier, C. (Eds) 2019. *A Tribute to the Reptiles and Amphibians of Australia and NewZealand*. Australian Herpetological Society. Reed New Holland.

Wilson, S., 2012. *Australian Lizards. A Natural History*. CSIRO Publishing.

Wilson, S., 2015. *A Field Guide to Reptiles of Queensland*. Second Edition. Reed New Holland.

Wilson, S. and Swan, G., 2008. *What Lizard is That? Introducing Australian Lizards*. Reed New Holland.

Wilson, S. and Swan, G., 2017. *A Complete Guide to Reptiles of Australia*. Fifth Edition. Reed New Holland.

GLOSSARY

Arboreal: Living in trees.
Depressed: In relation to build, dorsally flattened.
Dorsal: Relating to the upper surface.
Femoral pores: A row of pores under the thighs of some dragon lizards.
Frontoparietal scales: Paired shields, sometimes fused, on the middle of the head.
Gular fold: A skin fold across the throat, diagnostic in some dragon species.
Interparietal scale: Scale located centrally between parietal scales.
Labial scales: Scales along upper and lower lips.
Nape: Back of neck.
Nasal scale: Scale surrounding the nostril.
Ocelli: Ring-shaped spots, usually pale with dark centres.
Parietal scales: A pair of large scales on the rear of the head.
Postauricular fold: A skin fold behind the ear, diagnostic in some dragon species.
Preanal pores: Pores in front of the vent in some dragon lizards.
Reticulum: Netted pattern.
Scapular fold: A skin fold along the shoulder, diagnostic in some dragon species.
Spinifex: Grasses of the genus *Triodia*. Critical habitat for many desert lizards.
Subdigital lamellae: Enlarged modified scales under the digits, usually in reference to geckos.
Supraciliary scales: Row of scales immediately above eye.
SVL: Snout-vent-length. Standard measurement for a lizard excluding tail.
TL: Total length. Large species often measured to include tail.
Tubercles: Enlarged scale rising above the surrounding scales.
Ventral: Relating to the underside.

INDEX

A

Acritoscincus
    duperryi 65
Amalosia
    lesueurii 25
    rhombifer 25
Amphibolurus
    muricatus 144
Anepischetosia
    maccoyi 66
Anomalopus
    brevicollis 67
    verreauxii 68
Aprasia
    inaurita 55
    pulchella 55
Austroablepharus
    kinghorni 69

B

Bellatorias
    frerei 8, 70
    major 70
Blue-tongue
    Blotched 140
    Eastern 142
    Northern 142
    Pygmy 139
    Southern 140
    Western Slender 87
Bobtail 141
Bronzeback 60
Burton's Snake Lizard 59

C

Calyptotis
    scutirostrum 71
Calyptotis,
    Garden 71
Carinascincus
    ocellatus 72
    metallicus 64
Carlia
    longipes 73
    rhomboidalis 73
    rostralis 7
    rubigo 74
    tetradactyla 74
Carphodactylus
    levis 14
Chelosania
    brunnea 145
Chlamydosaurus
    kingii 146
Christinus
    marmoratus 45
Coeranoscincus
    reticulatus 75
Coggeria
    naufragus 76
Concinnia
    martini 77
Crenadactylus
    naso 26
    ocellatus 26
Cryptagama
    aurita 147
Cryptoblepharus
    buchananii 78
    juno 79
    pulcher 80
Ctenophorus
    adelaidensis 148
    aff. *spinodomus* 155
    cristatus 149
    decresii 150
    maculosus 151
    nuchalis 152
    ornatus 153
    pictus 154
    slateri 143
    vadnappa 156
Ctenotus
    greeri 81
    labillardieri 82
    pantherinus 83
    robustus 84
    striaticeps 85

Ctenotus
    Greer's 81
    Leopard 83
    Red-legged 82
    Robust Striped 84
    Stripe-headed 85
Cyclodomorphus
    casuarinae 86
    celatus 87
    gerrardii 88
Cyrtodactylus
    tuberculatus 46

D

Delma
    australis 56
    concinna 54
    impar 57
    tincta 58
    torquata 58
Delma
    Black-necked 58
    Collared 58
    Javelin 54
    Marble-faced 55
    Striped 57
Diplodactylus
    conspicillatus 27
    galeatus 28
    pulcher 29
    vittatus 30
Diporiphora
    amphiboluroides 157
    australis 158
    nobbi 159
    superba 160
Dragon
    Bearded, Central 168
    Bearded, Downs 167
    Bearded, Eastern 166
    Boyd's Forest 164
    Central Netted 152
    Chameleon 145
    Crested 149

Earless, Grassland 172
Earless, Smooth-snouted 171
Eyrean Earless 11
Goldfields Pebble-mimic 173
Gravel 147
Jacky 144
Lake Eyre 151
Long-nosed 161
Mallee Sand, Eastern 155
Mountain 169
Mulga 157
Nobbi 159
Ornate 153
Painted 154
Red-barred 156
Slater's Ring-tailed 143
Southern Angle-headed 164
Superb 160
Tawny 150
Water 4, 9, 162
Western Heath 148

Dtella
Crocodile-faced 48
Dubious 47
Northern Spotted Rock 47
Purnululu 44
Variegated 48

E

Egernia
cunninghami 89
kingii 90
rugosa 91
saxatilis 92
stokesii 93

Emoia
atrocostata 94

Eremiascincus
intermedius 95
isolepis 96

Eroticoscincus
graciloides 97

Eugongylus

rufescens 98

Eulamprus
quoyii 99
tympanum 12

F

Frill-neck 146

Gecko
Asian House 49
Beaded 32
Beaked, Brigalow 39
Beaked, Eyre Basin 39
Black Mountain 53
Broad-tailed 19
Broad-tailed, Mount Blackwood 18
Bynoe's 50
Cape York, Southern 52
Chameleon 14
Clawless, Northern 26
Clawless, South-western 26
Fat-tailed, Variable 27
Giant Cave, Northern 38
Giant Tree 37
Golden-tailed 43
Jewelled 41
Knob-tailed, Centralian 15
Knob-tailed, Prickly 15
Knob-tailed, Smooth 16
Leaf-tailed, McIlwraith 17
Leaf-tailed, Northern 20
Leaf-tailed, Southern 21
Marbled 45
Mesa 28
Mourning 51
Phasmid, Southern 42
Pretty 29
Ring-tailed, Cooktown 46
Sand-plain 32
Spiny-tailed, Northern 40

Stone, Eastern 30
Thick-tailed, Border 23
Thick-tailed, Common 22
Velvet, Fringe-toed 35
Velvet, Gulf Marbled 24
Velvet, Inland Marbled 34
Velvet, Lesueur's 25
Velvet, Robust 33
Velvet, Southern Spotted 36
Velvet, Western 31
Zigzag 25

G

Gehyra
dubia 47
ipsa 44
nana 47
variegata 48
xenopus 48

Glaphyromorphus
nigricaudis 100
punctulatus 101

Gnypetoscincus
queenslandiae 102

Goanna — see Monitor

Gowidon
longirostris 161

H

Harrisoniascincus
zia 103

Hemidactylus
frenatus 49

Hemiergis
peronii 104
talbingoensis 104

Hesperoedura
reticulata 31

Heteronotia
binoei 50

I

Intellagama
lesueurii 4, 9, 162

K

Karma

murrayi 105
Keeled Legless Lizard 62

L
Lampropholis
  guichenoti 106
  mirabilis 107
Land Mullet 70
Lashtail
  Horner's 163
  Swampland 170
Lepidodactylus
  lugubris 51
Lerista
  allanae 108
  ameles 108
  bipes 109
  bougainvillii 110
  microtis 111
  nichollsi 112
  parameles 64
  praepedita 113
Lialis
  burtonis 59
Liburnascincus
  mundivensis 114
  scirtetis 115
Liopholis
  inornata 116
  whitii 117
Lissolepis
  coventryi 118
  luctuosa 119
Lophognathus
  horneri 163
Lophosaurus
  boydii 164
  spinipes 164
Lucasium
  damaeum 32
  stenodactylus 32
Lygisaurus
  foliorum 120
  zuma 121

M
Magmellia
  luteilateralis 122
Menetia
  greyii 123
Moloch
  horridus 165
Monitor
  Black-headed 184
  Desert Pygmy 177
  Freckled 184
  Gould's 180
  Lace 174, 185
  Merten's Water 181
  Pygmy Mulga 179
  Sand 180
  Short-tailed Pygmy 176
  Spiny-tailed 175
  Spotted Tree 183
  Yellow-spotted 182
Morethia
  obscura 124
  ruficauda 125
Morethia
  Shrubland Pale-flecked 124

N
Nactus
  cheverti 52
  galgajuga 53
Nangura
  spinosa 126
Nebulifera
  robusta 33
Nephrurus
  asper 13
  amyae 15
  levis 16
Notoscincus
  ornatus 127

O
Oedura
  bella 24
  cincta 34
  filicipoda 35
  tryoni 36
Oligosoma
  lichenigera 128
Ophidiocephalus
  taeniatus 60
Ophioscincus

ophioscincus 129
Orraya occulta 17

P
Paradelma
  orientalis 61
Perentie 178
Phyllurus
  isis 18
  platyurus 19
Pletholax gracilis 62
Pogona
  barbata 166
  henrylawsoni 167
  vitticeps 168
Proablepharus
  reginae 130
Pseudemoia
  pagenstecheri 131
  spenceri 132
Pseudothecadactylus
  australis 37
  lindneri 38
Pygmaeascincus
  sadlieri 133
Pygopus
  lepidopodus 8, 63
  nigriceps 63
Rankinia
  diemensis 169
Rhynchoedura
  eyrensis 39
  mentalis 39

S
Saiphos
  equalis 134
Saltuarius
  cornutus 20
  swaini 21
Saproscincus
  challengeri 135
  czechurai 136
  mustelinus 137
Scaly-foot
  Brigalow 61
  Common 8, 63
  Western 63
Shingleback 141

Skink
  Bartle Frere 138
  Beech 103
  Black Mountain 115
  Black Rock 92
  Black-tailed Bar-lipped 100
  Brown Sheen 98
  Cunningham's 89
  Desert 116
  Dwarf, Common 123
  Earless, Eastern 104
  Earless, Peron's 104
  Elf 97
  Fine-spotted Mulch 101
  Fire-tailed 125
  Fraser Island Sand 76
  Gidgee 93
  Grass 106
  King's 90
  Litter, Sun-loving 121
  Litter, Tree-base 120
  Lord Howe 128
  Major 8, 70
  Mangrove 94
  Martin's 77
  McCoy's 66
  Metallic 64
  Mourning 119
  Murray's 105
  Nangur Spiny 126
  Norfolk Island 128
  Northern Bar-lipped 96
  Northern Narrow-banded 95
  Ocellated 72
  Orange-speckled Forest 122
  Outcrop Rock 114
  Pink-tongued 88
  Prickly Forest 102
  Pygmy, Magnetic Island 133
  Rainbow, Black-throated 7
  Rainbow, Blue-throated 73
  Rainbow, Closed-litter 73
  Rainbow, Orange-flanked 74
  Rainbow, Southern 74
  Saxicoline Sun 107
  Shade, Challenger 135
  Shade, Czechura's 136
  She-oak, Tasmanian 86
  Short-necked Worm 67
  Snake-eyed, Buchanan's 78
  Snake-eyed, Elegant 80
  Snake-eyed, Juno's 75
  Snake-eyed, Kinghorn's 69
  Snake-eyed, Ornate 127
  Snake-eyed, Spinifex 130
  Snake-tooth, Three-toed 75
  Spencer's 132
  Swamp 118
  Three-lined, Eastern 65
  Three-toed 134
  Tussock 131
  Verreaux's 68
  Water, Eastern 99
  Water, Southern 12
  Weasel 137
  White's 117
  Yakka 91
  Yolk-bellied Snake 129
Sleepy Lizard 141
Slider
  Bougainville's 110
  Chillagoe Fine-lined 64
  Inland Broad-striped 112
  Limbless Fine-lined 108
  Retro 66
  South Coast Five-toed 111
  West Coast Worm 113
  Western Two-toed 109
Strophurus
  ciliaris 40
  elderi 41
  jeanae 42
  taenicauda 43

T

Techmarscincus
  jigurru 138
Thorny Devil 165
Tiliqua
  adelaidensis 139
  nigrolutea 140
  rugosa 141
  scincoides 142
Tommy Round-head 158
Tropicagama
  temporalis 170
Tympanocryptis
  intima 171
  lineata 172
  pseudopsephos 173
  tetraporophora 11

U

Underwoodisaurus
  milii 22
Uvidicolus
  sphyrurus 23

V

Varanus
  acanthurus 175
  brevicauda 176
  eremius 177
  giganteus 178
  gilleni 179
  gouldii 180
  mertensi 181
  panoptes 182
  scalaris 183
  tristis 184
  varius 174, 185

W

Worm-lizard
  Granite 55
  Red-tailed 55

OTHER TITLES IN THE REED CONCISE GUIDES SERIES:

Reed Concise Guide: Animals of Australia
Ken Stepnell
ISBN 978 1 92151 754 9

Reed Concise Guide: Birds of Australia
Ken Stepnell
ISBN 978 1 92151 753 2

Reed Concise Guide: Frogs of Australia
Marion Anstis
ISBN 978 1 92151 790 7

Reed Concise Guide: Insects of Australia
Paul Zborowski
ISBN 978 1 92554 644 6

Reed Concise Guide: Snakes of Australia
Gerry Swan
ISBN 978 1 92151 789 1

Reed Concise Guide: Spiders of Australia
Volker W. Framenau and Melissa L. Thomas
ISBN 978 1 92554 603 3

Reed Concise Guide: Wild Flowers of Australia
Ken Stepnell
ISBN 978 1 92151 755 6

OTHER BOOKS ON HERPETOLOGY FROM REED NEW HOLLAND INCLUDE:

A Complete Guide to Reptiles of Australia (5th edition)
Steve Wilson and Gerry Swan **ISBN 978 1 92554 602 6**
A Field Guide to Reptiles of New South Wales (3rd edition)
Gerry Swan, Glenn Shea and Ross Sadlier **ISBN 978 1 92554 608 8**
A Field Guide to Reptiles of Queensland (2nd edition)
Steve Wilson **ISBN 978 1 92151 748 8**
Tadpoles and Frogs of Australia (2nd edition)
Marion Anstis **ISBN 978 1 92554 601 9**
A Tribute to the Reptiles and Amphibians of Australia and New Zealand
Australian Herpetological Society, edited by Chris Williams and Chelsea Maier

For details of these books and hundreds of other Natural History titles see:
www.newhollandpublishers.com
and follow ReedNewHolland on Facebook and Instagram